VISUALIZATION of RECEPTORS IN SITU:
Applications of Radioligand Binding

Methods in Visualization

Series Editor: Gérard Morel

VISUALIZATION of RECEPTORS IN SITU:
Applications of Radioligand Binding

Emmanuel Moyse, Ph.D.
Slavica M. Krantic, Ph.D.

CRC Press
Boca Raton London New York Washington, D.C.

Library of Congress Cataloging-in-Publication Data

Moyse, Emmanuel.
  Visualization of receptors in situ : applications of radioligand binding / Emmanuel
Moyse, Slavica M. Krantic.
    p.   cm. — (Methods in visualization)
  Includes index.
  ISBN 0-8493-0042-8 (alk. paper)
    1. Radioligand assay. I. Krantic, Slavica M. II. Title. III. Series.

QP519.9.R34 M69 2000
572'.33—dc21

00-060869
CIP

© 2001 by CRC Press LLC

No claim to original U.S. Government works
International Standard Book Number 0-8493-0042-8
Library of Congress Card Number 00-060869
Printed in the United States of America 1 2 3 4 5 6 7 8 9 0
Printed on acid-free paper

SERIES PREFACE

Visualizing molecules inside organisms, tissues, or cells continues to be an exciting challenge for cell biologists. With new discoveries in physics, chemistry, immunology, pharmacology, molecular biology, analytical methods, etc., limits and possibilities are expanded, not only for older visualizing methods (photonic and electronic microscopy), but also for more recent methods (confocal and scanning tunneling microscopy). These visualization techniques have gained so much in specificity and sensitivity that many researchers are considering expansion from in-tube to *in situ* experiments. The application potentials are expanding not only in pathology applications but also in more restricted applications such as tridimensional structural analysis or functional genomics.

This series addresses the need for information in this field by presenting theoretical and technical information on a wide variety of related subjects: *in situ* techniques, visualization of structures, localization and interaction of molecules, functional dynamism *in vitro* or *in vivo*.

The tasks involved in developing these methods often deter researchers and students from using them. To overcome this, the techniques are presented with supporting materials such as governing principles, sample preparation, data analysis, and carefully selected protocols. Additionally, at every step we insert guidelines, comments, and pointers on ways to increase sensitivity and specificity, as well as to reduce background noise. Consistent throughout this series is an original two-column presentation with conceptual schematics, synthesizing tables, and useful comments that help the user to quickly locate protocols and identify limits of specific protocols within the parameter being investigated.

The titles in this series are written by experts who provide to both newcomers and seasoned researchers a theoretical and practical approach to cellular biology and empower them with tools to develop or optimize protocols and to visualize their results. The series is useful to the experienced histologist as well as to the student confronting identification or analytical expression problems. It provides technical clues that could only be available through long-time research experience.

Gérard Morel, Ph.D.,
Series Editor

GENERAL INTRODUCTION

Receptors are the molecules that recognize chemical messengers and consequently trigger biological reactions in the recipient cells. First coined by Langley in 1879, this concept applies to endogenous signals such as hormones, neurotransmitters, embryonic inducers, cytokines, growth factors, and antigen-presenting complexes, and also to exogenous drugs modulating physiological effectors. Receptor function relies on reversible binding of a few specific chemicals called ligands that are not altered after release from the receptor. The amplitude of ligand-induced biological reaction and the amount of receptor–ligand complexes are similarly related to ligand concentration by sigmoid functions, which points to the underlying first-order chemical equilibrium:

$$R + L \leftrightarrow RL$$

Accurate determination of biochemical equilibrium parameters is crucial to the knowledge of drug action. Since Scatchard's biophysical formalization in 1949, this major aim of pharmacological science has been addressed extensively by analyzing the interaction of radioactively labeled ligands (radioligands) with fresh tissue homogenates. Such an approach, known as radioligand binding, allows functional characterization and quantification of biological receptors, while preparative biochemistry and molecular biology permit identification of corresponding proteins and/or genes.

Radioligand binding on tissue homogenates can, however, escape classical detection by scintillation counting when target cells are either not numerous enough or scattered among nonreactive cells; this occurs in heterogeneous organs such as the brain. Such limitation was overwhelmed in the late 1970s by transposition of radioligand binding on fresh tissue sections and subsequent autoradiography, which means detection of radioactive molecules in a sample by apposition to a photographic emulsion in the dark and by subsequent silver grain revelation. Autoradiography subserves two distinct *in situ* applications of radioligand binding: (1) a qualitative one, i.e., localization of radioligand-bound receptors within the tissue slice; and (2) a quantitative one, i.e., measure of bound radioligand, because the optical density of an autoradiographic image is a quasi-linear function of radioactive content (at least in a given range of concentration). Autoradiography quantification thus allows determination of receptor binding characteristics in tissue section areas instead of homogenates, which has considerably increased the resolution of the radioligand binding approach in the two last decades.

This book provides the theoretical bases of radioligand binding and describes the practical realization of the compromise between receptor binding kinetics and histology.

ACKNOWLEDGMENTS

We express our gratitude to Dr. Rémi Quirion (Douglas Hospital Research Center, Verdun, Quebec, Canada) and to Dr. Alain Beaudet (Montreal Neurological Institute, Montreal, Quebec, Canada) for the original receptor visualizations from their laboratories that they allowed us to use in this book. We also wish to acknowledge the considerable amount of training and advice that we received from both of them in this field.

This work was carried out in the framework of the European "Leonardo da Vinci" project (grant F/96/2/0958/PI/II.I.1.c/FPC), in association with Claude Bernard-Lyon I University (http://brise.ujf-grenoble.fr/LEONARDO).

CONTENTS

Chapter 1

Radioligands and Their Preparation

Contents

DEFINITION

A radioligand is a ligand made radioactive by isotopic labeling and used as a probe to study the receptor of interest. The chemical structure of the molecule chosen for isotopic labeling is identical or similar to that of the endogenous ligand for the relevant receptor. Synthetic agonists and antagonists of the endogenous ligands that do not naturally exist can also be used as radioactive probes.

The chemical structure of the ligand that can be isotopically labeled is one of the following:

- Biogenic amine
- Amino acid derivative
- Peptide and protein
- Nucleoside and nucleotide
- Steroid

They can be used to study the following types of receptors:

- G-protein coupled
- Tyrosine kinase
- Serine/threonine kinase
- Cytokine
- Growth factor
- Cell adhesion
- Ionotropic
- DNA-binding

↩ Ligands used for isotopic labeling are produced by chemical synthesis (i.e., purified, naturally occurring ligands are currently not used).

↩ Chemical modification of the synthetic ligand is generally performed to introduce the structural entities allowing isotopic labeling.

↩ The advantage of such synthetic analogs (agonists or antagonists) is that they often have better affinity for the relevant receptor and are less degradable than their natural counterpart ligands.

1.1 THE NATURE OF LIGANDS

Five major classes of ligands are used for the preparation of radioactive probes.

↩ Probe = radioligand (ligand + radioactive label)

5

1.1.1 Biogenic Amines

Biogenic amines are molecules containing the amine (NH_2) function. They include catecholamines and indolamines. The former are derived from the amino acid tyrosine (e.g., dopamine, norepinephrine) and contain a catechol nucleus (3,4-dihydroxylated benzene ring). Indolamines are derived from the amino acid tryptophan (e.g., serotonin); they have a five-membered ring containing nitrogen joined to the benzene ring.

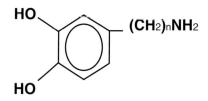

Figure 1.1 Generic structure of the catechol nucleus.

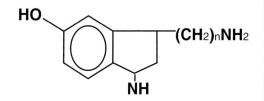

Figure 1.2 Generic structure of the indole nucleus.

1.1.2 Amino Acids and Derivatives

Amino acids are molecules containing a free carboxyl group and a free unsubstituted amino group on the α-carbon atom. These ligands include glutamic acid and γ-amino-butyric acid (GABA).

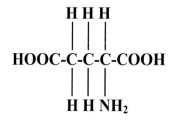

Figure 1.3 Glutamic acid.

```
        H  H  H
        |  |  |
HOOC-C-C-C-NH₂
        |  |  |
        H  H  H
```

Figure 1.4 γ-Amino-butyric acid.

1.1.3 Peptides and Proteins

This chemical category of ligands consists of molecules made from a few to a few hundred amino acids that are bound linearly by peptide bonds:

-(CH₂-CO-NH)-

1.1.4 Nucleosides and Nucleotides

Nucleosides are molecules containing a nitrogenous heterocyclic base and a sugar component, pentose. Nucleotides are nucleosides covalently bound to an additional molecule of phosphoric acid. The only known naturally occurring ligand in the nucleoside category is adenosine. It contains adenine as the purine base and D-ribose as the sugar component. Adenosine is recognized by different types of purinoceptors. Nucleotide ATP (adenosine containing three phosphate groups bound by a phosphodiester bond on the ribose 5′C atom) is the sole nucleotide that can be recognized by the same class of purinoceptors.

⇀ These bases are derivatives of either pyrimidines (uracil, thymine, and cytosine) or purines (adenine and guanine).
⇀ The pentose sugar is either ribose or 2′-deoxyribose.

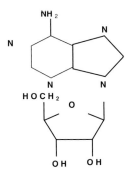

Figure 1.5 Adenosine.

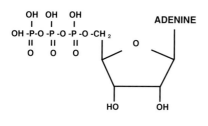

Figure 1.6 ATP.

1.1.5 Steroids

Steroids are derivatives of the saturated tetra-cyclic hydrocarbon perhydro-cyclopentano-phenanthrene. Steroids are ligands of intracel-lular receptors (cytoplasmic or nuclear). Bio-logically relevant steroid ligands are liposoluble hormones derived from cholesterol (sex steroids, mineralocorticoids, and gluco-corticoids). Hormones related to steroids func-tionally but not structurally, such as thyroid hormones (derived from tyrosine and iodine), are usually also considered as members of this ligand family.

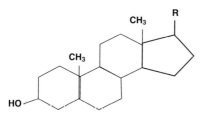

Figure 1.7 Cholesterol-related structure of steroids.

1.2 TYPES OF LABELS

The radioactive isotopes commonly used to label ligands for receptor visualization are:

- ^{3}H
- ^{35}S
- ^{14}C
- ^{125}I

☞ A radioactive isotope is an unstable atomic form of a chemical element.

^{3}H, ^{35}S, and ^{14}C are sources of β–radiation, whereas ^{125}I is a source of γ–radiation.

↪ Radiation is inherent to the instability of the isotope atomic nuclei. In the case of β–radiation for example:

$$^3H \rightarrow {}^2H + e^-$$

The choice of the isotope depends on the:

• chemical nature of the ligand
• position of the isotope's stable atomic counterpart in the ligand molecule
• method of labeling to be used.

1.2.1.   Chemical Characteristics of Commonly Used Isotopes

The intrinsic properties of a given isotope that should be considered prior to experimental setup are the following:

• Half-life

↪ Time period required for half of the radioactive atoms of a given isotope (independent of its amount) to return to their stable chemical form by emission of radiation

• Specific activity

↪ Gives the amount of radioactivity (reflecting the corresponding number of radioactive atoms) per defined unit of mass (e.g., mol) of the radionuclide. It is expressed in Ci/mmol

• Emission energy

↪ Energy produced during the radiation measured in MeV (mega-electron volt)

1.2.1.1 ^{3}H

Tritiated ligands provide excellent autoradiographic resolution (*see* Chapter 4). However, low emission energy of this isotope limits its application. In addition, ligands can be labeled by ^{3}H only during their chemical synthesis and are consequently purchased from commercial sources.

Biogenic amines, amino acid derivatives, steroids, and the nucleoside adenine are often labeled by substitution of stable hydrogen atoms in the ligand molecule with ^{3}H.

↪ The most frequently used

1. Characteristics:

• Half-life: 12.4 years
• Specific activity ≈ 25 to 130 Ci/mmole
• Emission energy: 0.018 MeV

↪ Long period
↪ Low activity
↪ Low energy

2. Advantages:

• More than one hydrogen atom can be substituted by the isotope in the ligand molecule.

⇨ *See* Figure 1.5.
⇨ To increase its specific activity
⇨ No modification of the chemical structure of the ligand

• Low emission energy; radioprotection easy

3. Disadvantages:

• Cannot be used for production of radioactive ligands in standard (research laboratory available) labeling procedures

⇨ Exclusively commercially purchased.

• Low emission energy; long half-life

⇨ Long time periods of exposition are necessary to reveal the presence of the labeled ligand–receptor complex by autoradiography.

1.2.1.2 ^{35}S

• This isotope is used to label nucleotide ligands such as ATP.

⇨ *See* Figure 1.6.
⇨ Rarely used

1. Characteristics:

• Half life: 87.4 days

⇨ Sufficiently long to allow autoradiography before radiolysis of the labeled ligand
⇨ Higher than that of ^{3}H
⇨ Relatively low, but higher than that of ^{3}H

• Specific activity: 3000 Ci/mole
• Emission energy: 0.167 MeV

2. Advantages:

• Relatively low emission energy

⇨ No important constraint for radioprotection
⇨ Autoradiography exposure time shorter than with ^{3}H

3. Disadvantages:

• Nucleotide structure is modified by substitution of one oxygen atom by ^{35}S in phosphate
• Risk of oxidation

⇨ *See* Figure 1.6.
⇨ Introduction of an unstable chemical bond
⇨ Necessity to use reducing agents (i.e., DTT, mercaptoethanol)
⇨ Exclusively commercially purchased

• Cannot be employed for radioligand production in standard (research laboratory available) labeling procedures

1.2.1.3 ^{14}C

^{14}C is used to label only a few ligands, such as diazepam (opioid receptor ligand), glutamate (ligand of ionotropic receptors), and steroid testosterone.

⇨ Rarely used

1. Characteristics:

- Half life: 5730 years
- Specific activity: 50 to 100 Ci/mole
- Emission energy: 0.156 MeV

⇝ Extremely long
⇝ Low activity, similar to that of ^{3}H
⇝ Relatively low, but higher than that of ^{3}H

2. Advantages:

- Low emission energy, equal to that of ^{35}S

⇝ No important constraint for radioprotection.

3. Disadvantages:

- Low emission energy, long half-life

⇝ Very long exposure time periods are necessary to reveal the presence of the labeled ligand–receptor complex by autoradiography.

- Cannot be used for production of radioactive ligands in standard (research laboratory available) labeling procedures

⇝ Exclusively commercially purchased

1.2.1.4 ^{125}I

^{125}I is used for labeling peptide and protein ligands, as well as for labeling thyroid hormones. It displays high emission energy and can be used to label ligands to a high specific activity and sufficiently long half-life.

⇝ Commonly used
⇝ Provides a good autoradiographic resolution

1. Characteristics:
- Half-life: 59.6 days

⇝ Long enough to allow the accomplishment of typical autoradiographic experiments before ^{125}I-labeled ligand radiolysis

- Specific activity: up to 2200 Ci/mmol

⇝ Theoretical specific activity of ligands labeled with ^{125}I equals that of ^{125}I-sodium salt : depends on the method used for radiolabeled ligand purification
⇝ ^{125}I provides much higher specific activities than other available isotopes.

- Emission energy:
  - 0.035 MeV γ-radiation
  - 0.032 MeV for X-rays

⇝ Relatively high
⇝ Photographic emulsions are activated by only a fraction of ^{125}I-emitted radiations: Auger electrons, energy of which is similar to ^{3}H-derived emissions
⇝ For autoradiographic applications, ^{125}I ensures resolution similar to ^{3}H.

2. Advantages:
- Simple labeling procedure

⇝ *See* Section 1.3.1
⇝ Feasible in research laboratories with reasonable radioprotection equipment

- High emission energy

⇝ Short autoradiographic exposure times

Radioligands and Their Preparation

• Excellent compromise between efficiency, feasibility, and cost of labeling

↪ Good resolution and sensitivity of autoradiographic labeling (*See* Chapter 4) are obtained with ^{125}I-labeled radioligands.
↪ Labeling procedures can be routinely performed in research laboratories.
↪ Low cost

   3. Disadvantages:

• Ligands to be labeled should naturally contain the benzene ring of tyrosine (e.g., thyroid hormones) or should be chemically modified to introduce this molecular structure.

↪ This limits the number ligands that can be labeled by ^{125}I.
↪ Chemical modification resulting from tyrosine introduction into the ligand molecule can alter the intrinsic properties (affinity, stability) of the ligand.
↪ However, some peptide and protein ligands can be labeled even in the absence of tyrosine by conjugation of free amino groups with ^{125}I-containing acylating agent (Bolton–Hunter reactive, *See* Section 1.3.2).

• High emission energy

↪ Labeling procedure requires particular radioprotection.

• Half-life moderately long

↪ Optimal utilization of iodinated ligands is limited to the first 15 days after labeling; they should not be used more than 30 days after iodination.

Table 1.1 Criteria for the Choice of Isotope for Ligand Radiolabeling

Type of Label	Labeling in Laboratory	Emission Energy	Radioprotection	Resolution	Availability
^{3}H	Not realizable	Very low	Minimal	Excellent	High
^{35}S	Not realizable	Low	Easy	Good	Limited
^{14}C	Not realizable	Low	Easy	Intermediate	Very limited
^{125}I	Rapid and easy	High	Important	Very good	Quite high

1.3   RADIOACTIVE LABELING PROCEDURES

The only radioactive procedure that can be performed under standard laboratory conditions is the labeling of peptides and proteins by ^{125}I isotope; it is generally called iodination. Iodination can be achieved by both chemical and enzymatic methods.

↩ Sodium salt of ^{125}I used for iodination is an extremely volatile molecule. All steps of iodination and purification (*see* Section 1.4) must therefore be performed in a confined space under a high-capacity extracting hood.

1. Chemical methods include:

• Substitution of ^{125}I into tyrosine residues in oxido-reducing reaction (Chloramine-T method)

↩ Applicable to peptides and proteins naturally containing, or chemically modified to introduce either tyrosine or histidine

• Conjugation of terminal amino groups by ^{125}I-containing acylating agent (Bolton–Hunter method)

↩ Applicable to peptides and proteins containing lysine residues.

2. Enzymatic method is based on:

• ^{125}I oxidation catalyzed by lactoperoxidase.

↩ Applicable to peptides and proteins naturally containing, or chemically modified to introduce tyrosine.

1.3.1   Chloramine-T Method

1.3.1.1   Principle

The principle of ligand iodination with this method is the oxidation of $^{125}I^-$ to $^{125}I^+$ by the oxidizing agent Chloramine-T (Chl-T), acting in this reaction as an acceptor of electrons:

↩ The most commonly used method of iodination

$$^{125}I^- \rightarrow {}^{125}I^+ + 2e^-$$

$$\text{Chl-T} + 2H^+ + 2e^- \rightarrow \text{Chl-T-H}_2 \text{ (reduced)}$$

Resulting in:

$$\text{Chl-T} + {}^{125}I^- + 2H^+ \rightarrow \text{Chl-T-H}_2 + {}^{125}I^+$$

↩ $^{125}I^+$ ions then allow an electrophilic substitution reaction on the benzene ring of tyrosine

$$^{125}I^+ + \text{Ligand} \rightarrow ({}^{125}I\text{-Ligand})^+$$

The reaction is stopped by addition of sodium metabisulfite.

↩ Reducing agent more potent than $^{125}I^-$ in Chl-T reduction.

13

1. Advantages:

- Rapid
- High yield
- Not expensive

↝ 5 min

2. Disadvantages:

- Risk of oxidative damage of ligand molecule due to its direct exposure to the high-energy ^{125}I isotope and strong oxidizing and reducing agents.

↝ Some proteins (e.g., calcitonin) are very sensitive to oxidative damage and cannot be radiolabeled by this technique even though they contain tyrosine.

1.3.1.2   Protocol

1. Materials and equipment:

- Hood in an area dedicated to ^{125}I manipulation

↝ All steps must be performed under the hood and behind lead protection blocks in areas dedicated to radioactive isotope manipulations.

2. Chemicals:

↝ Chloramine-T labeling kit is commercially available but its price is too high when the simplicity and the low price of included reagents are considered.

- ^{125}I in NaOH solution, 100 mCi/mL

↝ Purchased from standard commercial sources.

- NaH_2PO_4
- Na_2HPO_4
- Chloramine-T
- Sodium metabisulfite

↝ Synonym: sodium pyrosulfite

3. Solutions:

- Ligand containing tyrosine: 5 to 10 μg in 10 μL double-distilled water

↝ Peptide and protein aliquots can be prepared in advance and stored at $-20°C$ or $-80°C$. The absence of ligand proteolysis due to freezing/thawing should however be checked by HPLC before preparing the aliquot series. Small and relatively stable peptides when correctly preserved (at $-80°C$ without repeated thawing/freezing), can be used even up to a few years after aliquot preparation.

- Sodium phosphate buffer, 0.5 *M*, pH 7.5
- Chloramine-T, 1 mg/mL water
- Sodium metabisulfite, 1 mg/mL water
- Bovine serum albumin, 10% in water

- Trifluoroacetic acid, 0.1% in water

↝ **Must always be prepared fresh**
↝ **Must always be prepared fresh**
↝ Depending on the purification procedure, *see* Section 1.4
↝ Depending on the purification procedure, *see* Section 1.4

• Diluent buffer for iodinated ligand storage

↬ Depends on the nature of the iodinated ligand

↬ Buffers generally used are 0.1 *M* citrate (pH 6) and 0.2 *M* acetate (pH 4.6).

4. Procedure

• The reagents **must be added in the following order** and **mixed shortly after every addition**:

↬ Iodination procedure is carried out with low ^{125}I/peptide (protein) ratio so as to minimize radiation-associated damage and the number of ligand molecules that are substituted with more than one ^{125}I atom.

– Aqueous ligand solution	**10 µL**
– ^{125}I	**10 µL**
– Sodium phosphate buffer	**20 µL**
– Chloramine-T	**10 µL**

↬ 1 mCi

• Wait **40 sec**
• Then add
   – Sodium metabisulfite **20 µL**
• Wait 10 seconds, then add:
   – Bovine serum albumin **100 µL**

↬ For purification by ion-exchange chromatography

or

   – Trifluoroacetic acid **100 µL**
❑ *Following step*

↬ For purification by HPLC
↬ Proceed to the ligand purification immediately (*see* Section 1.4).

1.3.2 Bolton–Hunter Method

This method is based on stepwise lysine-containing peptide and protein ligand iodination without their direct exposure to oxidizing agents.

• Bolton–Hunter reactive (hydroxysuccinimide ester of hydroxyphenylpropionic acid) is labeled with ^{125}I by Chloramine-T method.

↬ First step of iodination

$$Chl\text{-}T + 2^{125}I^- + 4H^+ + 2[\text{succinimide-ester-}OH]^- \rightarrow Chl\text{-}T\text{-}H_2 + 2\,[^{125}I\text{-succinimide-ester}]^+ + 2H_2O$$

↬ The iodinated hydroxysuccinimide ester is separated from the reaction mixture by extraction into benzene and recovered by solvent evaporation under vacuum.

• Purified iodinated hydroxysuccinimide ester is then allowed to react with the free amino groups of lysine residues of peptides and proteins.

↬ Second step of iodination: the conjugation reaction

$$[^{125}I\text{-succinimide-ester}]^+ + [\text{protein}] \rightarrow [\text{succinimide-ester}] + [^{125}I\text{-protein}]^+$$

Iodinated peptides are purified by cation-exchange chromatography or by HPLC

☞ *See* Section 1.4.

1. Advantages:

• Ideal for labeling ligands susceptible to radiation and oxidative damage by either Chloramine-T (*see* Section 1.3.1) or Lactoperoxidase (*see* Section 1.3.3) method

☞ Bolton–Hunter method (in contrast to Chloramine-T and Lactoperoxidase methods) avoids direct exposure of peptides and proteins to potentially noxious ^{125}I solution and strong oxidizing and reducing agents.

• Labeling kit is commercially available

2. Disadvantages:

• Delicate if to be performed without the use of the labeling kit

☞ If to be performed without use of labeling kit, ^{125}I-labeled hydroxysuccinimide ester is very unstable under iodination reaction conditions: step 1 (from ^{125}I-sodium addition to ester extraction should be achieved in 20 seconds).

• Labeling kit is expensive

☞ Commercial kit is, however, time- and trouble-saving.

1.3.3 Lactoperoxidase Method

In this method, iodination is enzymatically catalyzed by lactoperoxidase.

1.3.3.1 Principle

The enzymatic method of labeling by ^{125}I is based on the lactoperoxidase-mediated iodination of tyrosil residues of peptides, and proteins in the presence of H_2O_2. The role of H_2O_2 is to form the reactive transition complex with the enzyme; the activated enzyme complex attacks $^{125}I^-$. Oxidized $^{125}I^+$ ions then allow an electrophilic substitution reaction on the benzene ring of tyrosine.

$$2\text{Peroxidase} + 2H_2O_2 \rightarrow 2[\text{Peroxidase-}H_2O_2]$$

$$2[\text{Peroxidase-}H_2O_2] + {}^{125}I^{2-} \rightarrow$$
$$2\text{Peroxidase} + 2H_2O + O_2{}^{125}I^+$$

$$^{125}I^+ + \text{Ligand} \rightarrow ({}^{125}I\text{-Ligand})^+$$

☞ Peroxidase of milk origin (lactoperoxidase) is very efficient in this reaction; for reasons that remain unknown, peroxidases from other sources (e.g., horseradish peroxidase) are inefficient.

1. Advantages:

• Mild chemical conditions of labeling, with low risk of oxidative damage

☞ Alternative method of tyrosine-containing ligand labeling

• No risk of damage due to exposure to reducing agents

↝ Simple tenfold dilution of the reaction mixture is sufficient to stop the enzyme action.

2. Disadvantages:

• Only tyrosine-containing ligands with highly accessible tyrosil residues are susceptible to iodination by this method

↝ This considerably limits the number of ligands that can be labeled by the Lactoperoxidase method.

• Risk of radiation-associated damage due to direct exposure of the ligand to ^{125}I

↝ Risk identical to that generated by Chloramine-T method

1.3.3.2 Protocol

1. Materials and equipment:

• Hood in an area confined to ^{125}I manipulation

↝ All steps must be performed under the hood and behind the lead protection blocks in the areas confined to radioactive isotope manipulations.

2. Chemicals:

↝ Lactoperoxidase labeling kit is commercially available but its use is not justified given the simplicity of the procedure and the availability and low price of the required reagents.

• ^{125}I in NaOH solution, 100 mCi/mL
• Lactoperoxidase
• NaH_2PO_4
• Na_2HPO_4
• H_2O_2

↝ Purchased from standard commercial sources

3. Solutions:

• Sodium phosphate buffer, 0.5 *M*, pH 7.5
• Lactoperoxidase: 2.5 to 25 mg/mL in phosphate buffer

↝ To be determined for each ligand

• Ligand containing tyrosine: 5 µg in 25 µL phosphate buffer

↝ Ligands stored in azide or sulfhydryl-containing solutions should be dialyzed prior to iodination because the latter compounds inhibit lactoperoxidase.

• H_2O_2, 0.88 m*M*

↝ To be prepared **fresh** as a water solution using 30% perhydrol.

4. Procedure:

↝ Iodination procedure is carried out with low ^{125}I:peptide (protein) ratio so as to minimize radiation-associated damage and the number of ligand molecules that are substituted with more than one ^{125}I atom.

• Add reagents **in the following order and mix shortly after every addition**

– ^{125}I stock solution	**8 – 15 µL**	↬ 0.5 to 1.8 mCi
– Ligand	**25 µL**	
– Lactoperoxidase	**1.5 µL**	
– H_2O_2	**1 µL**	
– Phosphate buffer	**to 50 µL**	
• Wait:	**2 sec**	
• Then add:		
– Phosphate buffer	**500 µL**	↬ To stop enzymatic reaction
❏ *Following step*		↬ Proceed to the ligand purification immediately (*see* Section 1.4).

1.4 PURIFICATION OF IODINATED LIGANDS

After iodination, the ligands are in the cationic form. They can therefore be purified from the reaction mixture by cation-exchange chromatography. This can be achieved technically either by:

↬ Purification is crucial in obtaining ligands iodinated to high specific activities.

• Cation-exchange chromatography
• High performance liquid chromatography (HPLC) using cation-exchange columns

The principle of the purification procedure is based on the separation of the mono-iodinated peptides or proteins from free iodine and unlabeled peptides or proteins. It can be divided into a few steps:

1. The cation-exchange column is first equilibrated with the appropriate buffer in a manner such that the resin pHi value is inferior to the ligand pHi value. The ligand is therefore adsorbed by the resin when the reaction mixture passes through the column.

2. Free $^{125}I^-$ anion is eliminated by simple addition of the elution buffer at low ionic strength (this anion is not retained by anionic resins such as would be the case with cation exchangers).

↬ To be sure that the iodinated ligand will not be eluted during this step, the ionic strength of the buffer should be about 100-fold lower than that used for elution of the iodinated ligand.

3. Iodinated and unlabeled ligands are then eluted with high ionic strength buffer. Iodinated ligands are retained by anionic resin more efficiently than unlabeled ligands (the presence of $^{125}I^+$ in the labeled ligand increases its electronegativity compared to the unlabeled ligand). Consequently, the labeled ligand is eluted after the unlabeled one.

4. The elution fractions are collected and an aliquot of each fraction is counted in a γ-counter to determine the elution profile.

The most radioactive fractions are preserved and tested on tissue sections to check the proportion of specific binding that can be measured with the relevant iodinated ligand.

↪ *See* Chapter 3.

1.4.1   Purification by Cation-Exchange Chromatography

1.4.1.1   Principle

The ion-exchangers commonly used to purify iodinated peptides and proteins are either resins derived from cellulose or gels derived from cross-linked dextran (Sephadex).

1. Advantages:

• Simple
• Not expensive

↪ No special equipment is needed.

2. Disadvantages:

• Low resolution of purification
• Purified ligands have relatively low specific activity

1.4.1.2   Protocol

1. Materials and equipment:

• Pump, fraction collector, column for chromatography (diameter 1 cm; height 25 cm), γ-radioactivity counter

↪ All steps must be performed under the hood and behind the lead protection blocks in areas confined to ^{125}I isotope manipulations.

2. Chemicals:

• CM-52 cellulose (Pharmacia)

↪ Should be activated according to manufacturer's instructions

• CH_3COONH_4

3. Solutions:

• 0.002 *M* CH_3COONH_4, pH 4.6
• 0.2 *M* CH_3COONH_4, pH 4.6

↪ Equilibration buffer
↪ Elution buffer

4. Procedure:

• Equilibrate the column with equilibration buffer.	**2–3 column volumes**

• Apply the iodinated sample to the top of the column and **allow to enter the column.** **100 μL**
 ↝ Volume of 10% bovine serum albumin (BSA)
 ↝ *See* Section 1.3.1.2.

• Rinse out the free iodine by passing the equilibration buffer through column. **20–25 mL**
 ↝ This buffer volume should be collected and divided into approximately 10 fractions (fraction size 2 mL).

• Elute unlabeled and iodinated ligand in: **60 mL**
 – 0.2 *M* acetate buffer, pH 4.6
 ↝ The elution volume should be collected and divided into 30 fractions (fraction size 2 mL).

– Take an aliquot of each fraction. **10 μL**
 ↝ Count in γ-radioactivity counter.

Represent graphically the radioactivity profile.
 ↝ *See* Figure 1.8. This profile should display two peaks, the first corresponding to the free iodine and the second corresponding to the labeled ligand.
 ↝ The **specific activity** of the iodinated ligand is calculated as a ratio of the radioactivity present in fractions corresponding to the second peak (R_2) over the sum of the radioactivity present in all fractions (total radioactivity of the eluate: R_T): R_2/R_T.
 ↝ They correspond to the second peak and its descending slope.
 ↝ Pool these fractions.

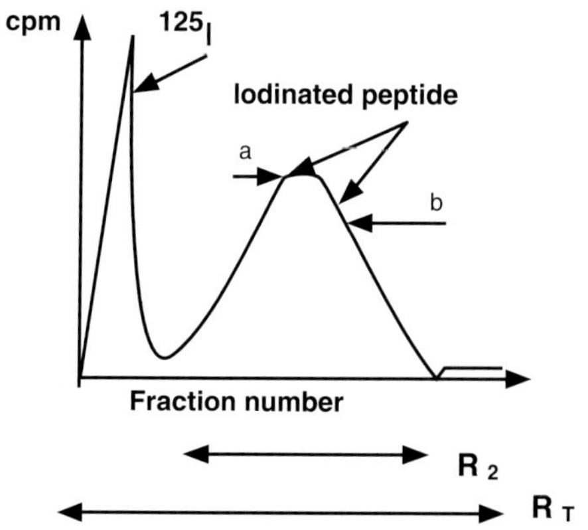

Preserve the most radioactive fractions.

Figure 1.8 Typical iodination profile after ion-exchange chromatography. Arrows a and b indicate the fractions to be kept.
 ↝ To have about 200,000 cpm per 100 μL

• Dilute the pooled fractions in:
 – 0.2 *M* acetate buffer, pH 4.6

• Prepare aliquots and keep frozen until use   **–20˚C**

↪ The aliquots are to be used not later than 30 days after iodination.

❏ *Following step*

↪ Perform receptor binding test with the purified ligand.

1.4.2   Purification by HPLC

1.4.2.1   Principle

The purification of iodinated ligands by HPLC is based on hydrophobic adsorption of peptides and proteins on silica gel. The silica gel column is polymerically bonded with different chemical functions, such as aliphatic carbon chains of different size, that allow hydrophobic interactions with peptides and proteins. The mixture to be separated is applied to the column in hydrophilic solution (stationary phase) in order to allow peptide/protein attachment to the column. The proportion of hydrophobic solution (mobile phase) is then increased, and peptides/proteins are eluted when the strength of hydrophobic interactions with the column's carbon chains becomes less important than that of the interactions with the mobile phase solution.

HPLC separation thus allows the separation of two or more peptides with equal molecular weight but different hydrophobicity. An important advantage of HPLC is the relatively short time necessary for separation. Indeed, this time is short in comparison to traditional chromatography, by applying high inlet pressure instead of atmospheric pressure.

HPLC purification comprises two steps. During the first step, the free $^{125}I^-$ anion is eliminated (prepurification). During the second step, the iodinated peptide is eluted (purification by HPLC, *strictu senso*).

1. Advantages:

• High degree of purification
• Rapid
• Purification of ligands to high specific activities

2. Disadvantages:

• Expensive equipment is needed

1.4.2.2 Protocol

1. Materials and equipment:

• HPLC display, sonicating bath, fraction collector, γ-radioactivity counter

• Sep-Pak column

⇨ For prepurification performed to remove the free iodine

• μ Bondapak C18 column

⇨ For purification (this column is an integral part of HPLC display)

2. Chemicals:

• Trifluoroacetic acid (TFA)
• Acetonitrile (ACN)
• Citric acid monohydrate

⇨ Stationary phase
⇨ Mobile phase

3. Solutions:

• 0.05% and 0.1% TFA in water
• 100% ACN
• 20, 25, 30, 50, and 75% ACN in 0.1% TFA
• 0.1 *M* citric acid buffer, pH 6

⇨ Sonicate to obtain gas-free solution.
⇨ Sonicate to obtain gas-free solution.

4. Procedure: prepurification on Sep-Pak column:

• Activate Sep-Pak.	**10 mL**	⇨ Corresponds to volume of 100% ACN
• Equilibrate the column.	**10 mL**	⇨ Corresponds to volume of 0.1% TFA
• Apply the crude iodination mixture.	**170 μL**	⇨ *See* Section 1.3.1.2.
• Elute successively with: – 20% – 25% – 30% ACN in 0.1% TFA	**500 μL**	⇨ Discard these eluates into ^{125}I waste.
• Elute successively with: – 50% – 75% ACN in 0.1% TFA	**500 μL**	⇨ Keep these eluates.
• Evaporate the eluate under a nitrogen stream to reduce the volume.	**200–300 μL**	⇨ The purpose of this step is to decrease, to a minimum, the ACN concentration.

5. Procedures: purification by HPLC:

• Set up HPLC display:
 – Flow rate   **1 mL/min**
 – Initial conditions   **0.05% TFA**

– Final conditions	**100% ACN**
– Separation time	**40 min**
• Set up the fraction collector	**0.5 min**
• Inject the sample.	**200–300 μL**
• Allow to run.	
• Take an aliquot of each fraction.	**10 μL**
• Represent graphically the radioactivity profile.	

↝ = 0.5 mL

↝ Count in γ-radioactivity counter

↝ *See* Figure 1.9.
↝ In contrast to purification by traditional ion chromatography, after the HPLC, purification only one peak of radioactivity is observed since the free iodine was eliminated during the prepurification procedure.

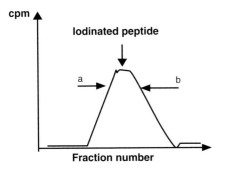

• Preserve and pool fractions corresponding to the peak.	**4–6 fractions**
• Dilute the pooled fractions in:	**100 μL**

Figure 1.9 Typical iodination profile after HPLC. Arrows a and b indicate the fractions to be kept.
↝ The specific activity of the iodinated ligand is approximated by the specific activity of ^{125}I sodium salt (i.e., 2200 Ci/mmol) used for iodination.
↝ To have about 200,000 cpm per 100 μL.

– 0.1 *M* citrate buffer, pH 6
• Prepare aliquots and keep frozen until use.   **–20°C**

– Allow to run.
– Dilute the pooled fractions in 0.1 *M* citrate buffer, pH 6, to have about 200,000 cpm per 100 μL.
– Prepare aliquots and keep frozen at –20°C until use.

❑ *Following step*

• Perform receptor binding test with the purified ligand.

↝ To have about 200,000 cpm per 100 μL.
↝ The specific activity of the iodinated ligand is approximated by the specific activity of ^{125}I sodium salt (i.e., 2200 Ci/mmol) used for iodination.
↝ The aliquots are to be used not later than 30 days after iodination.

↝ *See* Chapter 3.

Chapter 2

Frozen Tissue Preparation

CONTENTS

2.1   SAMPLING

2.1.1   Origin

↬ This step must obey the same demands as for biochemical homogenates while preserving sample morphology and cytology.

All types of biological material can be used, including:

- Organ
- Biopsy
- Cell culture
  - Plated monolayer
  - In suspension

- Cell smear

↬ For histological sections
↬ For histological sections
↬ *See* Figure 2.1.
↬ On presterilized glass coverslips
↬ In culture flask:
  •After cytocentrifugation
  •After pellet formation, freezing, and sectioning
↬ Cells stacked onto a glass slide with cyto-centrifuge (necessary in case of a low-density cell suspension).
↬ Preparation obtained by spreading a biological fluid: the technique of the smear yields a monocell layer on a glass slide.

2.1.2   Sample Obtention

2.1.2.1   Organs and biopsies

Dissection or biopsy handling must be performed as quickly as possible so as to minimize degradation of intrinsic proteins (including receptors) in unfixed tissue.

➥ Tiny pieces (i.e., less than 2 mm diameter), should be included in O.C.T. at room temperature before freezing.

Hollow organs (such as vertebrate olfactory organ) should be included in O.C.T. and kept under vacuum several minutes before freezing.

2.1.2.2   Cultured cells

2.1.2.2.1   USE OF PLATED CELLS

➥ Only in the case of adherent cells

Primary cultures of adherent cells can be directly processed for radioligand binding, provided that a flat and sterile support (like a coverslip) had been laid at the bottom of the culture box before plating the cells. Then the cell-bearing coverslip can be taken out of the culture medium and transferred to receptor labeling assay.

➥ The cell-bearing coverslip will then be processed exactly like a slide-mounted section (*see* Chapter 3).

2.1.2.2.2   PREPARATION OF CELL PELLET

➥ For subsequent freezing and sectioning

1 = Cell suspension:
•Cell cultures
•Fine needle biopsies
•Floating cells
2 = Preparation of a cell pellet:
•Centrifugation
3 = Preparation of cell spread:
•3A-Smear if cell density is high
•3B-Cytocentrifugation if cell density is low

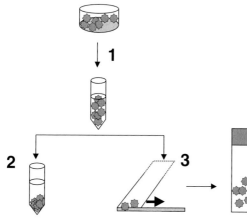

Figure 2.1 Cytological preparations.

2.2   FROZEN SECTIONS

Freezing is the transition of the water contained in cells and/or tissues from the liquid state to the solid state. This physical transformation produces sample hardening that allows cutting thin sections (5 to 30 μm) in a cryomicrotome.

2.2.1   Summary of the Procedure

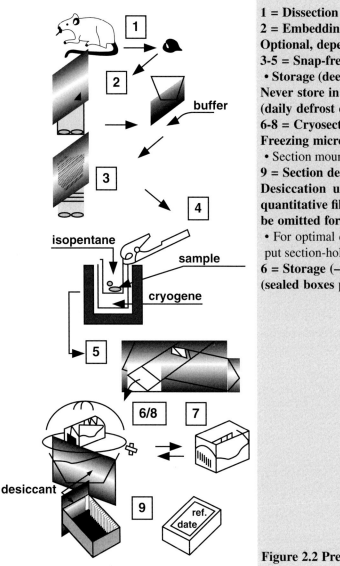

1 = Dissection
2 = Embedding in O.C.T. compound
Optional, depending on samples
3-5 = Snap-freezing(–45°C)
 • Storage (deep freezer, –80°C)
Never store in cryostat
(daily defrost cycle during the night)
6-8 = Cryosections:
Freezing microtome or cryostat
 • Section mounting on precoated slides
9 = Section desiccation and refreezing
Desiccation under vacuum at 4°C, for
quantitative film *autoradiography* only, to
be omitted for light microscopy purposes
 • For optimal cytological preservation, just
  put section-holding slides back in cryostat.
6 = Storage (–20°C or –80°C)
(sealed boxes provided with desiccant)

Figure 2.2 Preparation of frozen sections.

2.2.2   Materials / Reactants / Solutions

2.2.2.1   Equipment

• Jar for snap-freezing (–50 to –80°C)

↪ Allows rapid and homogeneous freezing
of tissue sample (–150 °C)
• Vacuum bell
↪ Desiccation of slide-mounted sections
• Freezer (–20°C or –80°C)
↪ Section storage

• Freezing microtome

2.2.2.2 Miscellaneous material

• Staining jars
• Pyrex beakers
• Tight-closing boxes (+ black tape)

• Dewar
• Steel forceps
• Histological razor, antiroll slide, brush, glass slides
• Slide holder or tray

☞ For deep-cold isopentane handling.
☞ For slide storage, before labeling and after dipping into liquid emulsion.
☞ For tissue freezing in liquid nitrogen.

☞ For instruments that must be kept clean by frequent wiping, either dry or with ethanol.
☞ Vertical position of slides facilitates section drying.

2.2.3 Sample Freezing

2.2.3.1 Principle

Freezing promotes blockage of all biological processes, including protein degradation, and tissue hardening.
Homogeneous freezing allows preservation of unfixed tissue cytology and thus cellular resolution.
Freezing homogeneity is best ensured by sample immersion in a fluid below 0°C ("snap-freezing").

☞ Frost-related hardness greatly facilitates sectioning without altering the molecular conformation of proteins.
☞ Sample size and choice of cryogenic agent are crucial.
☞ Temperature remains constant throughout freezing.

2.2.3.2 Cryogenic agents

• Liquid nitrogen (–196°C)

☞ **Danger** — burn risks.
☞ Use appropriate containers (i.e., not tightly closed (**blow-up risks**).
☞ Bad cryogenic agent for large and/or heterogeneous tissue samples because depth and speed of freezing promotes intra-sample cracking.
☞ Good cryogenic agent for biopsies collected in surgical units

• Dry ice (–78°C)

☞ Solid carbon dioxide; contact-induced freezing is usually heterogeneous and promotes morphological alterations

• Isopentane (methyl-2-butane) at –45°C (temperature adjusted by cooling isopentane jar in liquid nitrogen), followed by dry ice.

☞ Ideally suited for heterogeneous and thick samples (like rat brain)
☞ Immersion time must be optimized for each type of sample to avoid tissue cracking.
☞ Post-immersion drying without defrosting is required for isopentane to evaporate; otherwise viscous fluid remnants can damage sample periphery during subsequent storage.

• Propane (–45°C)   ⮑ Excellent cryogen. **Danger:** use carefully

2.2.3.3   Technique

1. Sample preparation:
Several methods are available for freezing,
including:

• Direct freezing   ⮑ Excess liquid at the surface of tissue can
form ice crust, which can be responsible for
tissue cracks. Drying promotes morphological alterations.

• Freezing after fixation and cryoprotection   ⮑ *See* Chapter 9.
• Freezing after embedding in O.C.T. compound   ⮑ Embedding in O.C.T. prevents cracking of
heterogeneous or morphologically tortuous
samples that are thus endowed with the physical resistance required for sectioning.
⮑ O.C.T. can be poured into the appropriate
mold.
⮑ For organs endowed with internal cavities,
O.C.T. embedding can be followed by a few minutes in a vacuum bell, so as to fill in internal
cavities with O.C.T. ("vacuum-embedding").

2. Freezing techniques:
Several techniques can be used; the best one is
to place the sample manually inside the cryogenic fluid from above (i.e., snap-freezing).

• In a liquid such as isopentane cooled in liquid
  nitrogen
   – Snap-freezing   **20–30 sec**   ⮑ To minimize cracking risks, the isopentane
                            at –45°C,   temperature and immersion time should be
                      then 15 min on   optimized for each type of biological sample.
                            dry ice   Present recipe is suitable for adult rodent brain.
                                             ⮑ Transfer onto dry ice for 15 min before
                                             wrapping allows isopentane remnants to evap-
                                             orate, thus avoiding peripheral deformation of
                                             sample.

• In liquid nitrogen   ⮑ Cracking risks for large samples (ice crystal
                                      formation in the case of muscle)
                                      ⮑ Well-suited for biopsy material

• On dry-ice surface   ⮑ Yields progressive freezing with a front par-
                                      allel to ice surface. Poor preservation of cytol-
                                      ogy above a small size threshold.

2.2.3.4   Storage

Frozen samples are stored in an air-tight wrap-   ⮑ Long-term storage at low temperature pro-
ping until use: motes some lyophilization, which will alter the
                                      binding capacity of receptors.

• Either at –196°C (liquid nitrogen)

☞ The lower the temperature, the better the storage. Because of crack risks, however, only biopsies or very tiny samples can be stored this way.

• Or in –20°C/–80°C freezers

☞ At –20°C, preservation of binding capacity is more or less limited in time (couple of months for μ-opioid receptors; couple of years for somatostatin receptors).

2.2.4 Realization of Cryosections

2.2.4.1 Material

A cryostat is a microtome placed in a powerful refrigerating compartment, wherein temperature can be adjusted and maintained down to –35°C, so that the tissue sample and razor can be held at optimal temperature for sectioning.
Razor is fixed; sample is mounted with O.C.T. on an object holder that can be moved in parallel to the razor.
Microtome movement can be either manual or motorized.

☞ Cryostat sections can be made from either pre-fixed or fresh-frozen tissues.
☞ Optimal sectioning temperature is lower for pre-fixed tissues (–25°C to –30°C) than for unfixed ones (–16°C to –20°C). The optimum also varies among tissues.

☞ Cryostats allow one to realize sections of 5 to 50 μm thickness.

1 = Storage of sample-holders
(rapid freezing device)
2 = Driving arm of sample holder
3 = Razor
4 = Razor holder
5 = Antiroll slide
6 = Tissue section
7 = Recuperation dish (tissue wastes)

Figure 2.3 Cryostat.

2.2.4.2 Glass slides for section collecting

For reliable adhesion of sections, histological glass slides must be precoated with a sticky material that is dried before using the slides.

☞ The precoating of slides is a particularly crucial step. Mind the following recipes.
☞ *Always use the same model of slides from one given supplier,* so as to avoid variations in slide thickness, which causes flaws on resulting film autoradiograms.

2.2.4.2.1 CLEANING OF SLIDES

1. Place slides in rack.
2. Immerse in household dish
   detergent.   **2–3 h**
3. Rinse under tap water.   **Abundantly**
4. Boil in distilled water.   **2 h**
5. Immerse in 95% ethanol.   **Overnight**
6. Rinse in distilled water.   **4 × 3 min**

☞ This step uses commercially available "pre-cleaned slides."

2.2.4.2.2   SUBBING

Prepare gelatin solution:

• 2% gelatin
• 0.05% chromium potassium sulfate

as follows:

1. Heat distilled water.	**70°C**
2. Drop gelatin powder into hot water.	**2%**
3. Dissolve using magnetic stirrer.	
4. Allow to cool.	**25–28°C**
5. Add chromium-potassium sulfate and allow to dissolve.	**0.05%**
6. Filter and avoid bubbles.	
7. Submerse slide racks in the gelatin solution.	**30 sec**
8. Dry out overnight in dust-free oven.	**at 37°C**
9. (Optional) dry subbed slides can be subbed again for adhesion improvement.	

↝ Gelatination is the most commonly used pre-treatment for radioligand binding on tissue sections.
↝ Adhesive compound
↝ Texture agent

↝ Use gelatin from porcine skin, type A, 375 Bloom (Sigma).

↝ Chromium potassium sulfate (Merck) is used as a texture agent; this compound is denatured by heat.

↝ Always use extemporaneously prepared solution.

↝ A double subbing will increase nonspecific binding of radioligands around sections.

2.2.4.2.3   STORAGE

In dust-free boxes for several months at 4°C.

2.2.4.3   Protocol

2.2.4.3.1   SAMPLE MOUNTING IN CRYOSTAT

1. Place object-holder directly in cryostat ($\approx$ –20°C) previously cooled to the appropriate temperature; its plate must be horizontal.

2. Place a drop of histological mounting medium (O.C.T.) on object-holder plate without bubbles.

3. Place frozen sample on the O.C.T. drop, so as to embed sample base within O.C.T.

4. Wait at least 15 minutes before sectioning to allow reliable hardening of O.C.T.

↝ **The sample must never thaw out**.

↝ Optimal temperature for cryostat sectioning varies among tissues:
• $\approx$ –16°C to –21°C
↝ Forceps must be used for depositing sample within the O.C.T. drop.
↝ For correct adhesion of object, O.C.T. must spread throughout the concentric rings of holder plate.
↝ One only has a few seconds before O.C.T. hardens (O.C.T. turns white) for the next step.
↝ Use pre-cooled forceps; maintain the sample in correct position for a few seconds, until O.C.T. hardens.
↝ Deficient O.C.T. embedding and/or hardening will cause sample to detach from holder when it hits the razor edge.

2.2.4.3.2 REALIZATION OF CRYOSECTIONS

1. Several parameters are involved in section quality, including:
   - Sectioning angle (12° to 15°)
   - Object size and shape

   ⇨ Fix it once and forever.
   ⇨ Sectioning is facilitated by decreasing object size and by making its razor-front side more convex.

   - Sample hardness
   - Temperature of object, razor, and anti-roll slide
   - Razor sharpness

   ⇨ Dependent on cryostat temperature
   ⇨ These temperatures must be homogeneous.
   ⇨ Any razor must be regularly sharpened; alternatively, use disposable blades.

   - Sectioning speed
   - Anti-roll position relative to razor edge

   ⇨ The speed of sample moving toward razor edge
   ⇨ This adjustment must be very accurate and must be watched throughout sectioning session.

   - Static electricity

   ⇨ This problem can often be solved by cleaning the razor and anti-roll with ethanol. It is aggravated by humidity.

2. Frozen object preparation:

   ⇨ Object holder should be placed on the microtome and locked in secure position.

   - Trim the O.C.T. block with a disposable razor blade into trapezoid shape, or at least avoid concavity facing razor front.

   ⇨ If sample is small, turn its largest side to the razor edge (*see* Figure 2.4).

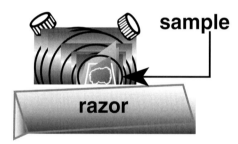

Figure 2.4 Realization of sections.

   - Adjust object-holder height relative to razor.

   ⇨ Sample surface must be flush with the razor.
   ⇨ Object surface and razor must be parallel:
   •A first parallelism between razor edge and inferior side of object
   •A second one between section plane and razor in order to obtain totality of sample on the section

3. Anti-roll slide adjustment relative to razor edge, i.e., parallelism and advancement

   ⇨ Anti-roll slide front edge must be parallel to razor edge.
   ⇨ Anti-roll slide front edge must overlay razor edge:
   - If anti-roll slide is advanced too far, section will carry cracks and/or folds parallel to razor edge (horizontal).
   - If anti-roll slide is not advanced enough, section will roll up or be torn vertically.

4. Section making:

• Reach the anatomical level of interest by trimming the sample to 25 to 30 μm thickness.

• Make sections at regular and sufficient frequency to ensure consistent thickness.

↪ The cryostat window must be kept closed as much as possible to avoid problems of temperature variation and condensation.
↪ Radioligand binding capacity is proportional to section thickness, the constancy of which determines the reliability of quantitative autoradiographic assays.
↪ Because of the object's tendency to compact in a cryostat atmosphere, a new section is thinner after a pause in slicing than after a series.

• Razor and anti-roll slide must be kept clean by frequent wiping with a brush or, when necessary, with an ethanol-dampened towel.

↪ Always move the brush or towel from bottom to the edge of razor — **never across the razor.**
↪ Wait for complete evaporation of ethanol before resuming slicing.

2.2.4.3.3   MOUNTING OF SECTIONS

For correct adhesion, tissue sections must defrost in contact with precoated glass slides. The newly cut section lies on the razor plane; this can be achieved either:
• With a warm slide (i.e., at room temperature) being positioned above the section, which is sucked up to the slide or
• With a cryostat-cold slide being laid down onto the section so as to draw it off the razor; defrosting is then provided locally by rubbing the back of the slide with your finger ("touch-mounting")

↪ Sections must be positioned in the distal third of the slide.

↪ If several sections are mounted on the same slide, make one single defrost for all.

↪ Adhesion and cytological preservation are excellent but section position is difficult to monitor and folds are difficult to avoid.
↪ To avoid bubbles and folds, a defrost front must be driven from one side of the section to the opposite side by rolling your finger underneath.
↪ When several sections are to be mounted together, the first one must not wait more than 10 minutes, to avoid desiccation.

2.2.4.3.4   REFREEZING OF SLIDE-MOUNTED SECTIONS

With desiccation

↪ Unfixed tissue sections cannot be dried at room temperature, to avoid cytology damage.

↪ This step improves adhesion in subsequent labelings, but hampers cytology preservation.

• At ambient air	**1–4 h at 4°C**	↪ In a plastic slide box laid on ice until the end of cryostat session (or slide box filling)
• In vacuum bell	**Overnight at 4°C**	↪ After 10 to 15 min coupling to a vacuum pump, the bell is closed and put in a fridge.

• Transfer to freezer

Without desiccation
• Direct freezing of new slides

⇔ Best preservation of cytology

2.2.4.3.5   STORAGE

In freezer   **–20˚C or –80˚C**

⇔ Slides must be kept in air-tight boxes.
⇔ Frozen tissue sections can be stored for months to years (depending on receptors studied).

Important: on the day of assay, the slides must defrost a few minutes at room temperature before any immersion.

⇔ If frozen slides are directly immersed, sections will fall off.
⇔ Wait just the minimal time for sections to dry out, by checking visually.

2.2.5   Other Preparations

Other classical histological procedures, like paraffin embedding and sectioning, cannot be used because of prefixation requirements.

❑ *Following step*

⇔ *See* Chapters 3 through 7.

Chapter 3

Radioligand Binding on Frozen Tissue Sections

Contents

3.1 PRINCIPLE

3.1.1 Aim

Radioligand binding on tissue sections is aimed at *in situ* visualization of the receptors that have been characterized biochemically with binding assays on tissue homogenates.

⇝ Incubation *per se* is directly transposable.
⇝ Tissue is much more difficult to rinse in sections than in homogenates.
⇝ Receptor labeling on tissue sections can be inhibited by endogenous ligands.

3.1.2 Receptor Kinetics Theory

Successful visualization of receptors by radioligand binding on tissue sections demands familiarity with the basic laws ruling chemical equilibriums, even for purely qualitative anatomical purposes.

⇝ For a more extensive review of receptor biochemistry and of pharmacological bases, the reader is referred to a comprehensive and didactic textbook: *Principles of Drug Action, 3rd edition*, W.B. Pratt & P. Taylor, Eds., Churchill Livingstone, 1990.

• Interaction of receptors R with their ligands L consists of a reversible complexation equilibrium:

$$R + L \leftrightarrow R\text{-}L$$

According to the law of mass action, concentrations of the equilibrium partners are related by the following equation:

$$[R].[L]/[RL] = K_d$$

K_d is the **dissociation constant** that characterizes the equilibrium. K_d is an affinity index: the higher the K_d, the lower the affinity of radioligand for the receptor.

⇝ Available experimental techniques allow one to monitor ligand concentration [L] and, provided that this ligand is radiolabeled, to measure the amount of complexed receptors [RL].

⇝ Biologically relevant parameters are K_d and the tissular concentration of binding sites.
⇝ Plot of bound radioligand vs. radioligand concentration ("free") yields a hyperbolic curve ending in a plateau (*see* Figure 3.1), corresponding to occupancy of all receptors present ("B_{max}"): saturation is reached. On such a "saturation plot," K_d is the concentration of radioligand ensuring one-half of the bound plateau value.

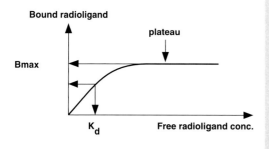

Figure 3.1 Receptor saturation by radioligand.

• For most hydrophilic hormones and neurotransmitters, K_d values are in the nanomolar range. B_{max} is commonly around 100 femtomoles per milligram protein.

↝ To compare, most enzymes display Michaelis constants (i.e., the parameter corresponding to the "dissociation constant" in enzyme/substrate interactions) in the micro molar range. Enzymes thus have 1000-fold less affinity for their substrates than receptors for their ligands.

• K_d and B_{max} are more accurately determined after transformation of direct saturation data into Scatchard coordinates. The Scatchard plot (B/F *vs.* B) is a descending straight line, the slope of which is $-1/K_d$ and abscissa of origin is B_{max}.

↝ This procedure arose from elimination of [R] between the law of mass action and the equation:

$$[R] = B_{max} - [RL]$$

↝ Hence the Scatchard equation:

$$B/F = -1/K_d \, (B - B_{max})$$

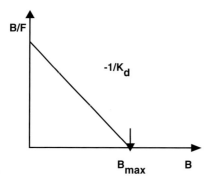

Figure 3.2 Scatchard plot of saturation data (for a single receptor type).

• In case of two different receptors for the same ligand in a given tissue sample, the Scatchard plot yields a curved line that can be resolved into two elementary straight slopes.

↝ The abscissa at the origin of the steepest line corresponds to the B_{max} value for the highest affinity binding site subtype (i.e., lowest K_d, highest slope absolute value). The abscissa at the origin of the second slope yields the B_{max} value for the lowest affinity binding site subtype (B_{max2}).

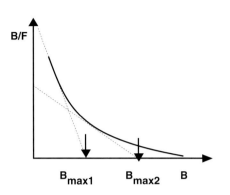

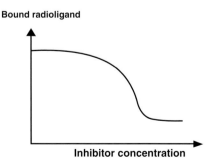

Figure 3.3 Scatchard plot of saturation data (for two receptor subtypes).

- Any given binding site can interact with a limited number of ligands, molecules of which share a common spatial determinant called the active site.

↪ This is the pharmacological **specificity** of the binding site.
↪ Specificity is thus a receptor parameter that is distinct from affinity. For example, among opiatergic drugs, DTLET is *the most specific* ligand of δ-opioid receptors (since it does not bind other opioid receptor subtypes, by opposition to most opiatergic drugs) while having *a very poor affinity* for δ receptors (K_d around 0.1 μM).
↪ Plot of bound radioligand *vs.* drug concentration yields a sigmoid curve spreading within 2 orders of magnitude.
↪ For any given drug, IC50 increases with the radioligand concentration used.

- Radioligand binding is inhibited by non-radioactive ligands of the same receptor through competition, in a dose-dependent manner.

The ability of a drug to compete for radioligand binding is quantified by the drug concentration ensuring half-maximal inhibition (**IC50**). IC50 can be used to calculate the inhibition constant K_i of the drug, according to the Cheng & Prusoff equation:

$$K_i = IC50 \times (1 + [L^*])/K_d$$

Where
$[L^*]$ = Radioligand concentration in assay
K_d = Dissociation constant of radioligand for assayed receptor

↪ K_i is a constant that is characteristic of the drug and independent of radioligand concentration.

Bound radioligand

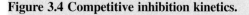

Inhibitor concentration

Figure 3.4 Competitive inhibition kinetics.

• Some competitive inhibitions of radioligand binding yield sigmoid curves spreading over more than 2 orders of magnitude (up to 4).

Such kinetics are indicative of more complex mechanisms than Michaelian interaction. This issue can be given statistical demonstration using the Hill plot:

↝ Shallow sigmoid curves
↝ This is not compatible with simple Michaelian kinetics.

$$\textbf{Log} \; (\textbf{B}_i/\textbf{B}_0 - \textbf{B}_i) = \textbf{N log [I]} - \textbf{N log IC50}$$

Where:

[I] = Concentration of competitor

B_i = Amount of bound radioligand in the presence of competitor at that concentration

B_0 = Amount of bound radioligand in the absence of competitor

N = Slope factor

The slope factor N is called the Hill number.

N = 1 indicates that the radioligand and the competitive inhibitor both bind to the same single class of non-interacting receptors.

N > 1 indicates positive cooperativity.

↝ The Hill plot also allows more accurate determination of IC50, and hence of drug K_i.
↝ Such data analysis can be performed with commercially available software.

↝ "Noncooperative" receptors

↝ Occupancy of one binding site by a ligand enhances the likelihood that other coupled sites on the same oligomeric receptor molecule will preferentially bind the same ligand or a related one.

N < 1 indicates either negative cooperativity or coexistence of multiple classes of non-interacting or non-interconvertible binding sites.

↝ Negative cooperativity occurs when a linkage between sites exists and occupation of one site decreases the likelihood that other sites on the same oligomeric molecule will bind the ligand.

• If different binding site subtypes exist for the same substance, the rank order of potency among binding drugs varies between binding site subtypes. This rank order of potency is thus a characteristic, called the **pharmacological profile,** of the receptor.

↝ The Hill coefficient can vary among a set of different competitors.
↝ Proper pharmacological characterization of a binding site must involve a sufficient number of different competitors, i.e., at least 6 to 8.

• Receptor/ligand interaction is also characterized by the time necessary to reach equilibrium.

Caution: all the above kinetic laws apply only "at equilibrium," i.e., once equilibrium is reached.

↝ This parameter is expressed as half-association time ($t_{1/2}$), i.e., time necessary to reach half-maximal binding level.
↝ Corresponding parameter is the half-dissociation time, i.e., the time required for the bound ligand to decrease to 50% of the equilibrium value.

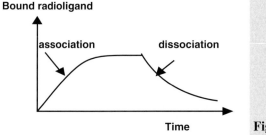

Figure 3.5 Association kinetics.

3.1.3   Radioligand Binding Sites

- All radioligands bind irreversibly to tissue components, yielding inescapable background or **nonspecific binding**.
- The parameter required for biochemical receptor analysis is called **specific binding** and cannot be measured directly.
- Specific binding can be inhibited by a "large excess" of nonradioactive ligand.
- The amount of radioligand bound to tissue preparation ("total binding") corresponds to the sum of two distinct components:

    – Specific, reversible binding to receptors
    – Nonspecific, irreversible binding to tissue

- Hence, the equation:

    Total = Specific + Nonspecific

➨ Nonspecific binding is directly proportional to radioligand concentration and cannot be inhibited by competition.
➨ Specific binding corresponds to the "bound" fraction of radioligand in the theoretical kinetic equations above.
➨ That is, 1000-fold the radioligand concentration
➨ The parameters directly accessible to experimental measure are:
- The sum of specific and nonspecific binding ("total binding")
- The nonspecific binding, i.e., bound radioligand remaining in the presence of 1000-fold-concentrated unlabeled competitor
➨ Specific binding can be calculated by subtraction between bound radioligand measures in two parallel assays on two identical tissue samples: one without ("total binding") and one with ("nonspecific binding") 1000-fold nonradioactive ligand.

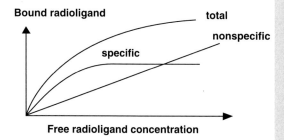

Figure 3.6 Components of specific binding.

3.1.4   Uses of Radiolabeled Sections

1. Radioactivity counting
Radiolabeled sections can be scraped off the slides, put in a clean assay tube, and transferred to a scintillation counter. This yields the global radioactivity content of the entire section.

➨ This application is convenient for preliminary test of radioligand binding.
➨ Receptive area of the section must be large enough so as not to dilute specific binding within background of nonreactive tissue.

2. Film ("dry") autoradiography
Radiolabeled sections are dried immediately after rinsing and then apposed onto a sheet of autoradiographic film.

➨ This is the most sensitive and most reliable procedure for binding site visualization and quantification. Resolution is limited to the macroscopic, subregional level.

3. Liquid emulsion ("wet") autoradiography
Radiolabeled sections are post-fixed, dehydrated/defatted, dipped into liquid emulsion and counterstained for light microscopic histology.

↪ This procedure is less sensitive than the "dry" one but yields cellular resolution.
↪ It requires cross-linking of radioligand to its specific binding sites, to avoid artifactual redistribution of specific labeling.

4. Coverslip-assisted autoradiography
Emulsion-precoated coverslips are glued by one edge onto dried-out radiolabeled sections, and are developed after exposure while remaining stuck to the section-bearing slide.

↪ This procedure was devised by Michael J. Kuhar in 1979. It yields cellular resolution with no more biochemical requirement than film autoradiography.
↪ It is very tricky and difficult to perform.

3.2 MATERIALS

• Bench-coating paper
• Clean slide racks
• Decontamination cuvette
• Disposable cytomailers
• Electric hair dryer
• Glass Coplin jars with vertical slots
• Glass staining dishes (250 mL) and slide racks
• Horizontal bench
• Tray with crushed ice for Coplin jars

↪ For slide drying
↪ For cleaning of nondisposable glassware
↪ For bath incubation with radioligand
↪ For slide drying
↪ For preincubation, rinsings, post-fixation
↪ For slide dehydration/defatting
↪ Suitable for handling radioactivity
↪ For rinsings and post-fixation

3.3 SLIDE DEFROSTING AND HANDLING

3.3.1 Slide Defrosting

3.3.1.1 Principle

Bring the slides from the freezer to room temperature and allow any moisture to vanish from tissue sections before dipping into first bath of assay.

↪ This step is crucial to avoid unfixed tissue sections separating from glass slides.

3.3.1.2 Protocol

1. Place the slides side by side, section upward, on a horizontal and clean substrate at room temperature.

↪ Do not allow any mechanical contact with sections; especially avoid letting the section side of slides touch the bench or dish walls or other slides.

2. Watch resorption of frost and moisture from tissue.

↪ Usually requires 2 to 3 minutes for slides taken out of a −20°C freezer, and 5 to 6 minutes out of −80°C freezer.

3. Process the sections **as soon as** they look dry, to avoid receptor-denaturing tissue tanning in air.

↪ Drying time must be kept similar for all slides of assay.

3.3.2 Slide Handling

- The slides must be put in, and drawn out of, each one of the successive baths with great care, at constant and low speed.
- **Do not agitate** any of the baths.
- The sections must never dry out.

↝ Avoid fluid flows over the tissue.

3.4 PREINCUBATION

This step is often (but not always) necessary to allow radioligand binding. It is believed to allow dissociation of endogenous ligand from tissular receptors, and should therefore be long enough for such dissociation to occur.

- One single buffered bath in receptor-dissociating conditions

❏ *Following step*

↝ Radioligand labeling of most hydrophilic substance receptors requires 15 to 30-minute preincubations.
↝ If no specific binding is obtained, the preincubation step must be added to the protocol or lengthened.
↝ Use the same buffer as for incubation with radioligand, but without the ions and cofactors required for optimal binding to receptors and without radioligand.
↝ *See* Chapter 5.

3.5 INCUBATION

3.5.1 Experimental Setup

Incubation of tissue sections with radioligand can be performed through either of the two following procedures.

1. Bath incubation
Slides are immersed in jars containing the incubation mixture.

↝ This method warrants constancy of mixture composition throughout incubation and reproducibility among all sections assayed.
↝ Disposable cyto-mailers (i.e., commercially available plastic boxes for 4 to 5 slides) allow one to minimize volumes of radioligand-containing mixtures.

2. Drop incubation
A drop of incubation mixture is deposited onto each section, the supporting slides being placed on a horizontal and clean substratum.

↝ This method is much less radioligand-consuming than the previous one.
↝ The drop must be prevented from evaporating, especially at room temperature, for example, by placing the slides in humid boxes throughout incubation.

The drop must be prevented from diffusing out of the tissue section. Horizontality of slides must be carefully watched and, in case preincubation is required, each slide must be wiped with absorbing paper all around each tissue section before depositing the drop.

Do not use the hydrophobic fabrics that are sold to encircle sections on glass slides for immunohistochemistry, because they might inhibit specific radioligand binding to receptors.

Figure 3.7 Drop incubation.

3.5.2 Incubation Medium

Must be prepared fresh on the day of assay

- **Ionic composition and pH** of assay buffer must be directly transposed from binding assays on tissue homogenates.
- **Isosmolarity** of assay buffer can be realized:
  - Either by increasing buffer concentration, which may reduce specific binding,
  - Or by adding sucrose, which usually has not much effect on specific binding,
  - Or by using cell culture media

Optimal binding is often obtained in widely hypotonic media, which compromises section preservation and adhesion.

Osmolarity of buffer is assessed with a picnometer.

TRIS buffer is isosmotic at 0.17 M.

Commonly used 50 mM TRIS buffer is made isosmotic by addition of 0.25 M sucrose.

- **Reducing nonspecific binding of radioligands** can be achieved by supplementing the assay buffer with:
- 2% Bovine serum albumin, or
- 0.1% Polyethylenimine
- **Additives protecting radioligands** from degradation

Nonspecific binding can also be reduced by some pretreatments, like 0.1 N HCl for GnRH (gonadotropin-releasing hormone) receptors.

Peptidic radioligands must be protected from degradation by tissular proteases by adding bacitracin (up to 0.1%).

Radioligands sensitive to oxidation (like aminergic molecules) are protected by the antioxidant 0.1% ascorbic acid (extemporaneously dissolved in assay buffer, the pH of which must be readjusted afterward).

– **Radioligand concentration** depends on the purpose of assay:

– For mere visualization purposes, it must be chosen at optimal specific over nonspecific (signal-to-noise) binding ratio, i.e., at K_d value or below.

– For competition assays (IC50 determinations), it must be fixed at K_d value.

– For saturation studies, a set of at least ten radioligand concentrations must be assayed in parallel, ranging from 0.1 to 10 predictive K_d value.

• **Ending incubation** of tissue sections with radioligand is realized merely by removing either the slides from the jar, or the drop from the slides.

❑ *Following step*

➯ It must be checked by scintillation counting on an aliquot of incubation medium.
➯ Receptor distributions are thus analyzed at nonsaturating conditions; the fraction of endogenous receptors labeled by the probe is postulated to be identical in all tissue components.
➯ On the basis of kinetic laws of chemical equilibria.
➯ K_d and B_{max} can be calculated by Scatchard transformation of competitive inhibition kinetics assayed at one single, nonsaturating concentration of radioligand. Such an approach will fail to reveal additional receptor subtype with lower affinity (higher K_d).

➯ *See* Chapter 6.

3.6 WASHING

Rinsing ensures elimination of remaining unbound radioligand under conditions that limit dissociation of receptor-bound molecules.

• Efficient rinsing of 20-µm-thick sections requires at least 2 consecutive baths of fresh buffer at 4°C, totaling a 10-minute period.

• The total washing time must be kept identical for all slides of assay.

• For subsequent autoradiography, slides must be quickly dipped through distilled water.

• For ligand-receptor systems reaching equilibrium in less than 15 minutes, the total rinsing duration must be decreased; radioligand labeling of glutamate receptors thus involves a 30-second rinse.

❑ *Following step*

➯ Rinsing of tissue sections is more difficult than rinsing of homogenates.

➯ The rinsing baths must be placed in ice-containing trays.
➯ The rinsing baths must be renewed with ice-cold fresh buffer every 20 slides.
➯ The rinsing duration determines the fraction of bound radioligand that dissociates.
➯ This precaution avoids salty and proteic deposits on the slides after drying, but can be harmful to tissue sections.
➯ Whatever the total duration, rinsing must go through at least two different consecutive baths.

➯ *See* Sections 3.8 and 3.9.

3.7   SCINTILLATION COUNTING OF SECTIONS

- Sections are taken off their slides, by one of the two following procedures:
  - Immediately after rinsing, wipe the section off the slide with a piece of clean, thick filter paper (e.g., GF/B Whatman).
  - After drying, scrape the section off the slide with a clean, new razor blade.

➡ Section must not have dried at all. This is the easiest and best procedure.

- Resulting tissue remnants are transferred to appropriate vial for scintillation counter, with or without scintillation fluid, depending on the isotope and the counter.

➡ Put one section per vial.

3.8   DRYING (FOR DRY AUTORADIOGRAPHY)

Slides are deposited on a clean slide tray for histology, section-bearing extremity at top, at room temperature.

➡ The slide tray must be radioactivity-free to avoid contamination of new slides.

Exposure to a current of clean air speeds the drying and helps in avoiding radioligand dissociation and redistribution.

➡ Place an electric hair dryer 1 m from the drying slides.

3.9   FIXATION (FOR WET AUTORADIOGRAPHY)

3.9.1   Principle

- This step is necessary to cross-link specifically bound radioligand and to avoid artifactual redistribution of label during subsequent dipping into alcohols, liquid emulsion, and photographic treatments.

➡ In addition, it provides histological fixation that improves tissue preservation for subsequent microscopic observations.

- It can be realized only for the radioligands, molecules of which contain the appropriate chemical moiety, by one of the two following procedures.

➡ Incidentally, wet autoradiography could be performed without fixation (nor dehydration/defatting) with slowly dissociating ligands, like ^{125}I-α-bungarotoxin for nicotinic receptors.

- Fixation is performed immediately after the last rinsing bath (*See* Chapter 6).

➡ The quick rinse in distilled water can be omitted.

3.9.2 Protocols

3.9.2.1 Immersion in 4% glutaraldehyde

For radioligands containing at least one free amine radical (-NH$_2$) within their molecule.

• Radiolabeled sections are immersed at 4°C for 30 minutes in a phosphate-buffered solution of 4% glutaraldehyde.

⟿ Cross-linking of bound radioligand onto its specific binding sites yields variable efficiencies among different receptors. Post-fixation efficiency varies from 50% (radioligands of μ-opioid receptors) to 90% (somatostatin receptor radioligand).
⟿ Post-fixation reliability must be checked in preliminary tests.

• Preparation of fixative
  – For 100 mL:

25% Glutaraldehyde solution	**16 mL**
0.4 M Sorensen buffer	**12.5 mL**
Distilled water	**71.5 mL**

⟿ Commercially available
⟿ *See* below.
⟿ Fill cylinder to 100 mL.

  – For 100 mL Sorensen buffer 0.4 M, pH 7.4:

NaH$_2$PO$_4$ · H$_2$O	**0.718 g**
Na$_2$HPO$_4$ · 12H$_2$O	**12.5 g**

3.9.2.2 UV light exposure

For radioligands containing one azido radical:
• Radiolabeled sections are photoirradiated while still in buffer with a 254-nm UV lamp (4 watts) for 10 minutes, at a 10-cm distance, at room temperature.

⟿ Efficiency of such post-fixation does not exceed 15%.
⟿ Commercial model

❏ *Following steps*

⟿ *See* Section 3.10.

3.10 DEHYDRATION/DEFATTING

3.10.1 Principle

Tissue defatting avoids lipidic quenching of radioactivity, which cannot be neglected with liquid emulsion because of grain thinness.
Prior to solvent treatment, tissue sections must be dehydrated to avoid morphological artifacts.
This step is commonly performed in staining dishes for conventional histology, by transferring radiolabeled slides in appropriate glass slide racks.

⟿ Various ethanol concentrations are prepared with distilled water.

⟿ All ethanol baths must be frequently renewed and discarded as radioactive waste.
⟿ These glass dishes are easy to decontaminate.

3.10.2 Protocol

1. Dehydration through increasing alcohols:
- Ethanol, 70%   **2 × 5 min**   ⇨ Loss of bound radioligand is maximal at this step.

- Ethanol, 95%   **2 × 5 min**

- Ethanol, 100%   **2 × 5 min**

2. Defatting in xylene:
- Xylene   **2 × 15 min**

3. Rehydration through decreasing alcohols:
- Ethanol, 100%   **2 × 5 min**
- Ethanol, 95%   **2 × 5 min**
- Ethanol, 70%   **2 × 5 min**
- Distilled water   **2 × 5 min**

4. Drying   ⇨ Overnight in 37°C oven
Slides are now ready for dipping in liquid emulsion.

❑ *Following steps*   ⇨ *See* Chapter 4.

Chapter 4

Light Microscopic Autoradio-graphy

Contents

4.1 PRINCIPLE

4.1.1 Definitions

Autoradiography is the detection of radioactive molecules within a tissue by:

• Affixing onto a photographic emulsion
• Forming a latent image through exposure in the dark
• Developing of the image as silver grains

The number of silver grains per unit area of image (i.e., **silver grain density**) is a direct function of radioactivity content in the underlying sample.

➣ Autoradiography relies on the property of light-sensitive photographic emulsions (silver halogenates) to be activated by radiation from radioactive molecules.

➣ Incidentally discovered by Marie Curie, this phenomenon led to wide applications in biology and to elaboration of the specifically devised "nuclear emulsions."

➣ Silver grain density provides a quantitative index of radioactivity in the underlying tissue, but only if the number of silver grains in the corresponding area is large enough.

➣ The relationship between radioactivity and silver grain density is empirical.

Silver grain density determines the physical parameter called **optical density** (OD). It can be measured through the proportion of light absorbed by the image from a source of known intensity, according to the law:

$$OD = -\log(I/I_0)$$

Where:
I_0 = Intensity of light incident to the image
I = Light intensity recovered beyond the image
Autoradiography can yield two types of information:

• Qualitative: localization of radioactive molecules within sample
• Quantitative: concentration of radioactivity in sample

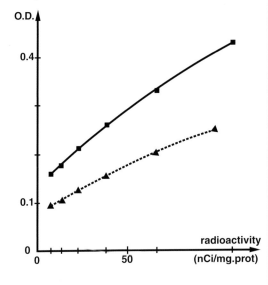

Radioligand binding sites can be visualized *in situ* at the light microscopic level by autoradiography of slide-mounted frozen tissue sections that have been radiolabeled as described in Chapter 3.

↪ Optical density depends on both the light source and the characteristics of the light-meter apparatus.
↪ Optical density is expressed in arbitrary units.

↪ The "semiquantitative" approach is limited to relative comparison between labeling intensities.
↪ Absolute quantification of autoradiographic labeling (to be expressed in radioactivity units) requires a calibration curve realized with standards of known radioactivities that have been autoradiographed together with the sections to quantify.
↪ This calibration curve depends on the:
•Radioactive isotope
•Type of emulsion
•Development protocol
•Duration of exposure

Figure 4.1 Relation between sample radioactivity and autoradiogram optical density.

↪ This is only one among numerous applications of autoradiography.

4.1.2   Mechanism

The basic mechanism of autoradiography is identical to that of photography.

↪ An image is created in photosensitive emulsion by incident energy from radioactive disintegrations in the dark instead of light radiations.

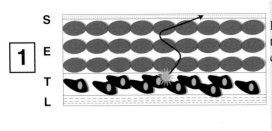

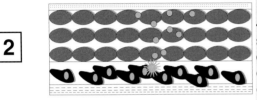

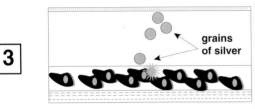

1 = Radiation
Bound radioligand contains radioactive isotopes that emit radiation characterizing the chemical element used.
• **S** = Film substratum
• **E** = Emulsion
• **T** = Tissue or cells
• **L** = Slide

2 = Exposure
↬ A "latent image" is formed within the emulsion affixed to radiolabeled sections. This process consists in Ag^+ ion activation, i.e., peripheral electron jump to a superior energy level that renders Ag^+ sensitive to chemical reduction to metallic Ag. This phenomenon occurs within the silver bromide crystal, in the absence of water. It merely results in a change of Ag^+ oxido-reductive potential without any visible effect.

3 = Development
↬ Latent image is transformed into silver grains by immersion in a weak reductant (called a developer); the image is then stabilized by a strong oxidant (called a fixer) that neutralizes developer and dissolves out unreacted emulsion grains.

$$Ag^+ + e^- \rightarrow Ag$$

Figure 4.2 Principle of autoradiography.

4.1.3   Characteristics

4.1.3.1 Isotopes used

Receptor ligands are usually radiolabeled with either ^3H, ^{125}I, ^{35}S, or ^{14}C.

↬ Various emulsions, called "nuclear emulsions," have been optimized for the various isotopes.

4.1.3.2   Photographic substrates

Nuclear emulsions for autoradiography exist basically as photographic emulsions in silver bromide salts diluted with gelatin. Their characteristics include:

• Grain size

• Emulsion layer thickness

↬ The wider the grain, the higher the sensitivity and the weaker the resolution.
↬ The thinner the grain, the lower the sensitivity and the higher the resolution.

• Support: either liquid as a gel (for tissue coating) or industrially spread over a transparent support (film)

↪ Distinction between macro- and micro-autoradiography

4.1.3.3 Exposure

The radiolabeled sample is kept in direct contact with the emulsion in absolute darkness during the time necessary for latent image formation (a few hours to several months).

↪ Exposure time is determined empirically and depends on tissular radioactivity concentration, radioisotope, and nuclear emulsion used.

4.1.3.4 Revelation (development)

Transformation of latent image into silver grains involves chemical processing similar to photographic development.

↪ The chemical protocols provided by manufacturers must be strictly and consistently followed. Image quality cannot be improved at this step of the autoradiographic procedure.
↪ Most emulsions, unlike photographic papers, are very sensitive to mechanical contact throughout developing.
↪ Each emulsion fits with one given developer (to be checked in manufacturer's notice). Use of inappropriate developer can fail to yield an image.
↪ Fixers are less diversified than developers, but must nevertheless fit the developer (*see* manufacturer's notice).
↪ Abundant rinsing is important to avoid subsequent damage to autoradiograms by oxido-reductant precipitates.

• Development (i.e., action of a chemical reductor called "developer" that reduces latent image into silver grains)
• Fixation (i.e., action of a chemical strong oxidant called "fixer" that neutralizes developer and dissolves out unreacted emulsion)
• Rinsing under tap water (>15 minutes) and distilled water (quick passage)

4.1.3.5 Efficiency

A 15% efficiency for an autoradiographic process is generally considered acceptable. A large fraction of latent images escapes reduction into silver during development.

↪ Knowing that:
• Radiation from radioactive isotope is emitted in three dimensions, whereas emulsion is present on one plane of section.
• Impression efficiency of emulsion is not 100%.
• Development process is optimal but not absolute.
• Background can arise from both section labeling and emulsion processing.
↪ This efficiency has not improved over many years, but autoradiography remains a highly sensitive detection method.
• **E = Emulsion**
• **C = Cells**
• **S = Slide**

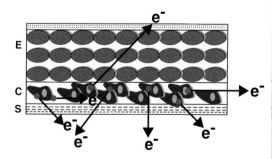

Figure 4.3 Autoradiographic efficiency.

4.1.3.6   Localization

Autoradiographic localization of bound radioligand molecules depends on the following parameters:
- Radiation energy, which depends on isotope and probe specific activity
- Emulsion grain size (the smaller the grains, the more accurate the localization)
- Emulsion thickness

⇝ At the light-microscopic level, the radioactive source position can be assimilated to the position of the grain.
⇝ If exposure time is too long, the latent image can extend around the radioactive source, which decreases the localization accuracy.

4.1.3.7   Detection

Autoradiographic signal can be detected:
- At the macroscopic level on film
- At the microscopic level using liquid, fine-grained emulsion

⇝ Both procedures can allow quantification of bound radioactivity, by ensuring appropriate exposure time and calibration with radioactive standards.

4.2   DRY AUTORADIOGRAPHY

4.2.1   Film Autoradiography

⇝ Macroautoradiography

4.2.1.1   Principle

Macroautoradiography consists of generating an image that will be physically separated from the biological sample. It uses industrially manufactured films with high sensitivity and weak resolution.

⇝ This type of radio-sensitive surface provides homogeneity and reproducibility and is thus especially appropriate for quantitative purposes.
⇝ Emulsion grain size is wider than mean cell diameter.

4.2.1.2   Choice of film

Different types of autoradiographic films are commercially available, each with specific characteristics, by Amersham/Pharmacia and Sigma.

⇝ Choice of film depends on the isotope used, the tissular concentration of radioactive molecules to detect, the levels of sensitivity, and the resolution desired.

1. Monoface/double-sided films

Commercially available films are coated with emulsion either on both faces (Hyperfilm MP) or on one single side (BioMax MR, Hyperfilm ^{3}H, Hyperfilm β-max).
Double-sided films have bigger-sized grains and greater sensitivity than monoface ones, but provide lower resolution.

⇝ Emulsion side of monoface films must be in direct contact with the radiolabeled sections.

⇝ Monoface films are more expensive than double-sided ones.

2. Preferential isotope sensitivity

Below are given, as examples the sensitivities of Amersham films for receptor studies:

• Hyperfilm MP	^3H, ^{125}I, ^{35}S, ^{14}C

↝ Maximum speed and sensitivity, lowest resolution

• BioMax MR	^{35}S, ^{14}C

↝ Sensitivity is increased (×2) by tabular shape of emulsion grains.

• Hyperfilm β-max	^{125}I, ^{35}S, ^{14}C

↝ Emulsion layer is covered by a protective anti-scratch layer that seriously impedes or totally blocks the weak beta particles of ^3H and the low energy Auger electrons of ^{125}I decay.

• Hyperfilm ^3H	^3H, ^{125}I

↝ Because there is no protective layer, care must be taken when handling this film to avoid damaging the emulsion and creating artifacts.

3. Grain thickness

The thicker the grain of emulsion, the higher the sensitivity and the lower the resolution.

↝ The film with thinnest grain, and thus with highest resolution, is Hyperfilm ^3H (Amersham).

4.2.1.3 Summary of the different steps

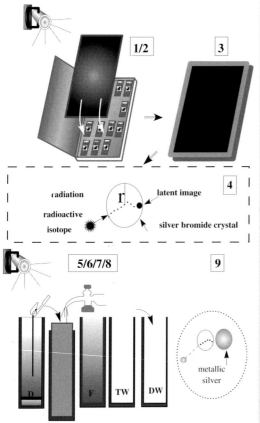

1 = Placing slides in cassette
2 = Film apposition
3 = Cassette storage
(ambient temperature)
4 = Macroautoradiographic exposure
5 = Development
6 = Rinsing under tap water
7 = Fixation
8 = Rinsings
 • Tap water (TW)
 • Distilled water (DW)
9 = Signal evaluation

Figure 4.4 Protocol for macroautoradiography.

4.2.1.4 Exposure

Film is affixed in the darkroom onto sections in a large-enough, lightproof lead cassette.
Emulsion must be directly in contact with radiolabeled tissue sections. Any air-space between section surface and emulsion will result in a flawed image.

↬ **Under inactinic light**
↬ Films must be handled only in the darkroom, under appropriate inactinic light (check manufacturer's instructions):
- Sodium light for Hyperfilm ^{3}H
- Dark brown light (Kodak O filter) for Hyperfilm MP, Hyperfilm β-max, BioMax MR

↬ Lightproofness must be checked regularly.
↬ Film must be pressed onto slides when cassette is closed; if necessary, add enough paper sheets to fill cassette.
↬ Slides must be placed side by side, sections up, so as to create a perfectly flat surface; avoid laying sections over neighbors.

4.2.1.5 Development

Two developer/fixer combinations are widely used:

- Kodak D19 and Kodak rapid fixer (undiluted stock solutions)
- X-ray film processing reagents; for example Kodak LX-24 developer and Kodak AL-4 fixer (both diluted 1:4).

↬ Under inactinic light
↬ Same as above (*see* Section 2.1.4)

↬ Do not combine heterologous reagents.

↬ For Hyperfilm ^{3}H

↬ For BioMax MR, Hyperfilm MP, and Hyperfilm β-max

4.2.1.6 Tissue counterstaining

After film removal, sections can be stained for conventional histology.

↬ Tissue preservation, although hampered by *in vitro* incubations without prefixation, is still good enough for subregional (supracellular) localization of autoradiographic labeling.

4.2.1.7 Film storage and handling

Films must be kept in:
- Dry
- Dust-free

storage, which protects them from mechanical contacts (especially Hyperfilm ^{3}H).

↬ Commercial transparent pockets are perfectly suited.

4.2.2 Coverslip Autoradiography

An emulsion-coated coverslip is stuck to the radiolabeled section.

↬ *W.S.Young and M.J. Kuhar, 1979, A new method for receptor autoradiography: (^{3}H) opioid receptors in rat brain, Brain Res. 179, 255–265.*
↬ Emulsion-coated coverslips are not commercially available and must therefore be prepared in the laboratory.

4.2.2.1 Materials

- Absorbent paper
- Aluminum foil
- Black tape
- Chinaware spoons (4)   ↝ For emulsion handling and homogenizing
- Clean coverslips
- Clean glass slides
- Dry-rite paper bags
- Instant glue   ↝ Loctite, Zyan-acrylate
- Jolly jars   ↝ Jolly jars are specially manufactured, oval-section cylinders that minimize the amount of emulsion required to cover the slides.
- Lightproof slide boxes
- NTB-2 (Kodak) emulsion
- Scale-rule
- Thermometer
- Thermostated and agitated water bath   ↝ Must accommodate jars vertically

4.2.2.2 Dipping the coverslips

↝ *Under inactinic light*

1. Before entering darkroom: ↝ Switch on warming bath 1 h before starting.

- On the Jolly jar, place two labels with permanent marker: ↝ Marker color must be visible under inactinic light.
    - One corresponding to the water volume necessary for emulsion dilution
    - The other corresponding to total volume (emulsion + water).
- Fill the Jolly jar with distilled water up to the lower mark.

2. In the darkroom, with appropriate inactinic light, open up emulsion container and:

- Pick up emulsion **gently** with china spoon.   ↝ Avoid mechanical stress to stock emulsion, which can increase auto-impression and thus background.
- Add as many spoons as required to raise water surface to full level.
- Transfer to warming bath.   ↝ Emulsion must not be touched with bare fingers nor with metallic tools.

3. 15 minutes later:

- Plunge a clean spoon into emulsion jar.   ↝ Make **slow** motions, to avoid creating bubbles.
- Stir gently, by moving the spoon 15 vertical turns in one direction and 15 turns in the other direction.   ↝ Separate used spoons for washing.

4. Repeat step 3 two more times at 15-minute intervals.   ↝ The entire process must take 1 hour. Increasing preparation time will increase background.

5. One hour after step 2, emulsion is homogeneous and ready to use. Remove bubbles by dipping 4 to 6 clean test slides before the actual ones to be autoradiographed.

⇨ Check the test slides against inactinic light, and make sure there are no bubbles.
⇨ Number of coverslips must be limited so that dipping duration does not exceed 1/2 h after the end of emulsion preparation.

6. Hold coverslips one-by-one with Brussel forceps, vertically.

- Dip into emulsion jar with a gentle and continuous movement, down to the bottom and back up.
- Still holding the coverslip vertically, gently blot it on a cushion of absorbent paper.

⇨ The coverslip must be dipped so as to leave the top 2 mm free of emulsion, for gluing onto the slide.
⇨ Emulsion-coated surfaces must be spared any mechanical contact, even when dried.

7. Place coated coverslips sub-vertically on a hotplate at 40°C.

Allow to dry overnight in the dark, before use.

⇨ Hotplate must be covered with a flat sheet of absorbent paper, to protect the heating surface without making the coverslips fall down.

4.2.2.3 Assembly procedure

⇨ ***Under inactinic light***

1. Use the "good" side of emulsionized coverslip toward the section.
2. Put one little drop of glue on the slide (not on coverslip), on the side of section opposite the slide top, where coverslip edge will be lifted up with a wooden match for development.

⇨ The glue sites (on slide and on coverslip) must be free of fingerprints and emulsion.
⇨ Place the glue drop at mid-width of slide.

3. Immediately apply emulsionized coverslip, so that its side opposite to the glued one protrudes 1/10 mm over the slide extremity.
4. Directly transfer the slide to exposure box.

⇨ Quadruplicate each type of autoradiographed section, for both development hazards and different exposure times.

4.2.2.4 Exposure

⇨ In absolute darkness at 4°C

1. After assembly of emulsionized coverslips, store the slides in lightproof slide boxes along with a dry-rite bag (one per box), close the box, fix lid with a circular band of black tape, and wrap the sealed box in aluminum foil.

⇨ Dry-rite bag can be fixed within the box by interposition of an additional clean, non-autoradiographed slide.

2. Exposure time is 5 to 6 times longer than exposure time on film.

⇨ Split the slides into at least two sets, to be developed after different exposure times. Usually 2 to 3 weeks if film autoradiography of this material needs 3 days; exposure duration of the second set must be determined after having developed the first one, on a quasi-proportional basis.

4.2.2.5 Development

1. Prepare as many bevelled wooden matches as there are slides to develop.
2. On each autoradiographed slide, lift up the free edge of coverslip with a bevelled wooden match.
3. Protocol:

• Developer   **90 sec**
(Kodak Dektol for NTB-2)   **at 17˚C**

• Distilled water   **dip, 4˚C**
• Fixer (rapid fixer, Kodak)   **10 min, 4˚C**
• Rinse in distilled water   **2 × 4min, 4˚C**
❏ *Following steps*
4. Counterstain and dehydration/mounting

➪ *Under inactinic light*

➪ Meanwhile, put photographic reagents to cool down in ice-box in darkroom.
➪ Manipulate delicately to avoid breaking coverslips.

➪ Cool from ambient temperature by transferring jar into ice-box; follow up while stirring and take out of ice when thermometer indicates 18˚C.

➪ *See* Chapter 5 for stain choice and protocols.
➪ Make sure that coverslip glue will not be dissolved by any of the staining steps.

4.3 WET AUTORADIOGRAPHY

4.3.1 Materials

• Absorbent paper
• Aluminum foil
• Black tape
• Chinaware spoons (4)   ➪ For emulsion handling and homogenizing
• Dry-rite paper bags
• Glass slides
• Jolly jars   ➪ Jolly jars are specially manufactured, oval-section cylinders that minimize the amount of emulsion required to cover the slides.
• Lightproof slide boxes
• Scale rule
• Slide rack (clean)   ➪ Free of any previous contact with radio-activity, even decontaminated
• Thermostated and agitated water bath   ➪ Must accommodate jars vertically

4.3.2 Procedure

4.3.2.1 Principle

Dipping of radiolabeled sections into liquid emulsion generates autoradiographic silver grains stuck directly on histological substratum.

➪ Fixation and staining of tissue sections after radioligand binding and autoradiography allows one to localize binding sites at the cellular level.

4.3.2.2   Choice of emulsion

Several liquid nuclear emulsions are commercially available, each one convenient for all four commonly used isotopes:

• Ilford K5	**0.2 μm**	↝ High sensitivity
• Kodak NTB-2	**0.26 μm**	↝ Moderate sensitivity
• Amersham LM-1	**0.25 μm**	↝ Moderate sensitivity

4.3.2.3   Summary of the different steps

↝ *Under inactinic light*

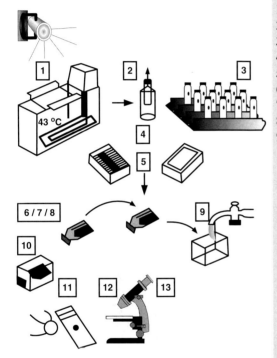

1 = Melting of emulsion (41°C)
2 = Dipping of slides
3 = Drying (4 hr – overnight)
4 = Storage in the dark (4°C)
5 = Exposure time
6 = Development (17°C)
7 = Rinsing
8 = Fixation (4°C)
9 = Rinsings
 • Tap water
 • Distilled water
10 = Dehydration
 • Graded alcohols
11 = Mounting (xylene, resin)
12 = Observation
13 = Interpretation (signal analysis)

Figure 4.5 Protocol for microautoradiography.

4.3.2.4   Dipping protocol

↝ *Under inactinic light* (Sodium light for the three emulsions cited)

1. Before entering darkroom:

• On the Jolly jar, place two labels with permanent marker: one corresponding to the water volume necessary for emulsion dilution, the other corresponding to total volume (emulsion + water).

 – Ilford K5   **1:1**

↝ Marker color must be visible under inactinic light.
↝ If more than 30 slides are to be coated, prepare an additional jar (Jolly or Borrel) for refills.
↝ Emulsion/distilled water 1:1 (vol/vol)

– Kodak NTB-2   **1:1**
– Amersham LM-1   **2:1**

↪ Emulsion/distilled water 1:1 (vol/vol)
↪ Emulsion/distilled water 2:1 (vol/vol)

• Fill the Jolly jar with distilled water to the lower mark.

2. In the darkroom, with **appropriate** inactinic light:

↪ Avoid mechanical stress to stock emulsion, which can increase auto-impression and thus background.
↪ Emulsion must not be touched with bare fingers nor with metallic tools.

• Open the emulsion container and pick up emulsion **gently** with china spoon.
• Add as many spoons as required to raise water surface to full level.
• Transfer to thermostated bath.

3. 15 minutes later:

• Plunge a clean spoon into emulsion jar.
• Stir gently, by moving the spoon 15 vertical turns in one direction and 15 turns in the opposite direction.

↪ Make **slow** motions to avoid creating bubbles.
↪ Separate used spoons for washing.
↪ If two emulsion jars are to be prepared, repeat each step for the second jar immediately after the first one, so that both jars are ready to use together and display the same properties.

4. Repeat step 3 two more times at 15-minute intervals.

5. 1 hour after step 2, emulsion is homogeneous and ready to use. Remove bubbles by dipping 4 to 6 clean test slides before the ones to autoradiograph. Do it again for refill.

↪ The entire dilution/warming process must take 1 h. Increasing preparation time will increase background.
↪ Check the test slides against inactinic light and make sure there are no bubbles left. Slide number must be limited so that dipping duration does not exceed 1 h after the end of emulsion preparation.

6. Emulsion coating is performed by immersing slides one-by-one into emulsion jar with a gentle and continuous movement, down to the bottom and back up.

↪ Emulsion must reach 0.5 cm above the last section, because of thickness gradient from slide bottom (*see* Figure 4.6).
↪ Slides can be processed two-by-two if they are held back-to-back; but they must be put to dry separately.
↪ Emulsion coating must be spared from any mechanical contact, even once dried.
↪ Do not knock the slides to avoid stressing emulsion and thus raise autoradiographic background.

Holding slides vertically, gently dab emulsion excess on a cushion of absorbent paper.

7. Place coated slides vertically (emulsion-coated tip down) on a clean slide rack. Allow to dry in ambient air, 4 hours, in the dark.

↪ Racks must be clean, and especially devoid of any radioactive contamination.

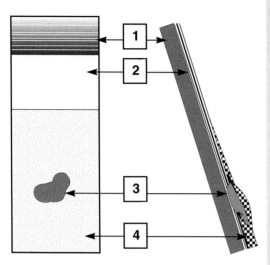

1 = Glass slide
2 = Slide precoating (gelatin)
3 = Tissue section
4 = Emulsion

Figure 4.6 Drying of emulsion-coated slides.

4.3.2.5 Exposure

4 hours (at least) after dipping:
1. Store emulsion-coated slides in lightproof slide boxes along with a dry-rite bag (one per box).
2. Close the box; fix lid with a circular band of black tape.
3. Wrap the sealed box with aluminum foil.
• Exposure time is 5 to 6 times longer than exposure time on film.

⇝ *In absolute darkness at 4°C*

⇝ Avoid any mechanical contact with emulsion-coating.
⇝ Dry-rite bag can be fixed within the box by interposition of an additional clean, non-autoradiographed slide.

⇝ Split the slides into several sets to be developed at different exposure times (three sets are ideal; adjust the two longer exposures on the basis of the first one).

4.3.2.6 Development

The protocol, as in general photography, is a routine one and must be kept strictly constant. It is never subject to experimental optimization, since its features have already been optimized by emulsion manufacturers.
• Developer (D19 Kodak)   **4 min**
                                        17°C

• Distilled water   **Dip**
                                        4°C
• Fixer (rapid fixer, Kodak)   **10 min**
                                        4°C

⇝ *Under inactinic light*

⇝ Set up water and fixer baths in ice-box 0.5 h before beginning.
⇝ Label the different baths (developer, fixer, water) with permanent marker, visible under inactinic light
⇝ Cool from ambient temperature by transferring jar into ice-box; follow up while stirring and take out of ice when thermometer indicates 18°C (due to glass thickness, liquid temperature will drop down 1 more degree).

• Rinse under tap water   **30 min**
• Rinse in distilled water   **Dip**
• Dry slides in a dust-free place

4.3.2.7   Tissue counterstaining   ⮑ *See* Chapter 5.

Chapter 5
Counter-staining

Contents

5.1 AIM

Counterstaining aims to visualize tissular structures without altering autoradiographic labeling.

↝ Coloration must be light enough so as not to mask the autoradiographic signal.

5.2 MATERIAL

Sections to be stained in the context of receptor autoradiography have not been fixed according to state-of-the-art prerequisites. This material can, however, be sufficiently stained for reliable identification of tissular and even cytological substratum.

↝ Histological stainings have been optimized and described in textbooks for paraffin sections of tissues prefixed with mostly paraformaldehyde-based mixtures.

5.2.1 Emulsion-Dipped Sections

5.2.1.1 Principle

The pretreatments these sections have gone through, do not forbid histological staining:
• Incubations in aqueous solutions above 0°C while totally unfixed, for at least 1 h
• Post-fixation in 4% glutaraldehyde
• Dehydration/defatting/rehydration
• Dipping in liquid emulsion
• Incubations in photographic reagents

↝ At least with the stains listed below

↝ *See* Chapters 3 and 4.

↝ Strong oxido-reducing agents

5.2.1.2 Procedure

Because these sections have already been fixed, they can be directly transferred to staining. However, their atypical pretreatments require optimization tests by reference to protocols of histological textbooks.

5.2.2 Film-Removed Fresh Sections

These sections have gone through:

↝ Fine cytology cannot be assessed.

- Incubations in aqueous solutions while totally unfixed   **Above 0°C 1 h at least**
- Kept unfixed   **At room temperature Days to weeks**

↝ Some stains will fulfill their tinctorial properties on these sections as such.

It is therefore highly recommended to fix them prior to staining, with the 4% glutaraldehyde method used for wet autoradiography of peptidic radioligands (*see* protocol in Chapter 4.3)

↝ Optional
↝ Powerful fixative

5.3 CHOICE OF STAINS

Numerous stains are classically used in light microscopic histology, each with elective tinctorial specificity toward some of the tissue components.

↝ The stain to use must be chosen based on the tissular elements best suited for localizing the investigated receptor type.

5.3.1 Nucleic stains

↝ Tinctorial affinity for nucleic acids

- Hemalun of Masson
- Cresyl violet

↝ Stains nuclei in dark blue/black
↝ Stains nuclei of all cells and Nissl bodies of neurons (hence cytoplasm of neuronal cell bodies); the tool for Nissl staining of nervous tissue

5.3.2 Cytoplasmic stains

- Eosin
- Toluidine blue

↝ Stains cytoplasmic elements in pink
↝ Stains cytoplasmic elements in light blue

5.3.3 Matrix stains

- Azan

- Mallory

↝ Azan stains:
- Collagen, bone, and cartilage in blue
- Mucus bluish
- Nuclei and muscle fibers in red
- Elastin in faint pink
- Erythrocytes and myelin in orange-yellow

5.4   PROTOCOLS

Each of the following protocols must be initially optimized to:

1. The biological material. Reaction of autoradiographed sections to the stain varies according to:
• Nature of the tissue
• Section thickness
• Previous binding and autoradiography steps
2. The bench, in particular the quality of distilled water used

A few autoradiographed sections must be used, one-by-one, to fix the following parameters:
• Duration of staining step (the thicker the section, the longer the staining)
• When necessary, amount of additive (acid for example) to staining and/or differentiating solutions
• Duration of differentiating steps

5.4.1   Hematoxylin-Eosin

5.4.1.1   Preparation of stock stain solution

5.4.1.1.1   HEMATOXYLIN

1. Dissolve:
• Potassium alun   **5 g**
• Distilled water   **100 mL**
2. Boil.
3. Add:
• Hematein   **0.2 g**
4. Allow to boil.   **A few minutes more**
5. Cool.
6. Filter.
7. Add:
• Pure acetic acid   **2 mL**

5.4.1.1.2   EOSIN

1. Dissolve:
• Eosin   **1 g**
• Distilled water   **100 mL**

☞ Only the most commonly used stains are considered below; the following selection should allow for correct localization of autoradiographic labeling in any organ.

☞ Staining properties of tissues are altered by wet autoradiography more than by dry, as compared to classical properties of paraffin-embedded prefixed samples.

☞ Stable 1 month at room temperature in dark bottles.

5.4.1.2   Protocol

1. Stain in hematoxylin.	**2 – 10 min**
2. Rinse rapidly in water.	
3. Differentiate in:	
• Ethanol, 95%	**99 vol**
• HCl	**1 vol**

☞ Shake the slide rack inside the bath constantly and watch color evolution; transfer to water as soon as color looks fine (to be checked under the microscope).
☞ Until nuclei take on a dark-blue color.

4. Rinse under flowing water.	
5. Immerse in eosin.	**1 min**
6. Differentiate in 70% ethanol.	

☞ Watch to determine when to stop this step. Shake constantly.

7. Further dehydrate in:	
• Ethanol 95%	**3 min**
• Absolute ethanol	**3 min**
8. Transfer to xylene.	**5 min**
9. Mount in resin.	

☞ At least

5.4.2   Mallory–Haidenhain

5.4.2.1   Preparation of stock stain solution

• Distilled water	**200 mL**
• Phosphotungstic acid	**1 g**
• Orange G	**2 g**
• Aniline blue	**1 g**
• Acid fuchsin	**3 g**

☞ Add the components one-by-one in the order listed. For each one, wait until complete dissolution before adding the next.

5.4.2.2   Protocol

1. Immerse slides 5 min in the stain solution.
2. Rinse under tap water a few seconds.
3. Dehydrate in graded ethanols (70%, 95%, 100%, 3 min each).
4. Transfer to xylene for at least 5 min.
5. Mount in resin.

5.4.3   Neutral Red

• Immerse slides 2 to 3 min in neutral red stock solution.
• Rinse in distilled water.
• Dehydrate in graded ethanols (70, 95, 100%) 1.5 min each.
• Transfer to xylene for 5 min.
• Mount in resin.

5.4.4 Cresyl Violet (Nissl Stain)

5.4.4.1 Preparation of stock stain solution

1. Dissolve:
• Cresyl violet **1.25 g**
• Distilled water **250 mL**
2. Filter.
3. Add:
• 10% Acetic acid **1.5 mL**

⇝ Can be stored several months at 4°C; must be filtered from time to time
⇝ Stir on magnetic agitator overnight.

⇝ For 250 mL
⇝ This proportion deserves optimization.

5.4.4.2 Protocol

1. Immerse slides in stain solution. **3 min**
2. Rinse in water bath. **15 sec**
3. Differentiate in 250 mL 70% ethanol supplemented with 15 drops 10% acetic acid. **1 – 3 min**
4. Further differentiate in 95% ethanol. **3 min**
5. Further differentiate in 100% ethanol. **3 min**
6. Transfer to xylene. **5 min**
7. Mount in resin.

⇝ Constantly shake slide rack inside the 70% ethanol bath; remove when sections look mid-dark violet.
⇝ Shake only from time to time.
⇝ Shake only from time to time.

Chapter 6

Double Labeling

Contents

6.1   THE IMMUNOHISTOCHEMICAL REACTION

6.1.1   Principle

The principle of the immunohistochemical reaction is to create a high-affinity complex between a tissular antigen and an exogenous antibody (called *primary antibody*).

⤳ Antigen/antibody complexation is the basis of humoral immune reactions in mammalian organisms. It is an equilibrium: K_d (*see* Chapter 3) is around 10^{-12} *M*. Such high affinity makes the complexation slow (12 h to reach saturation), but macroscopically irreversible in ambient laboratory conditions.

⤳ Several primary antibodies are now commercially available for identification of cell phenotypes.

The primary antibody:

1. Has been raised specifically against the antigen
2. Is applied *in vitro* onto the tissular section

⤳ Usually with tissue prefixation constraints

3. Is then visualized by coupling to:
• A chromogen chemical (fluorescent radical, enzyme)
• A colloidal gold particle

This coupling can be direct (chromogen is covalently linked to primary antibody) or, more often, indirect, which involves an antibody (called a *secondary antibody*) raised against immunoglobulins of the animal species used for generating the primary antibody.

⤳ Indirect coupling provides amplification-based increase of detection sensitivity.

6.1.2   Requirement for Tissue Prefixation

Antibody binding to tissular antigens generally requires prefixation of tissue.

Because radioligand binding to the receptors is abolished by tissue prefixation, this step must be performed after binding and autoradiography.

The fixative mixture depends on the primary antibody used. The most common, almost standard, fixative is 4% paraformaldehyde in pH 7.4 sodium phosphate buffer.
Some antibodies prove unable to react on post-fixed tissue sections.
For wet autoradiography combined with immunohistochemistry, a compromise is often worked out between the requirements of:
• Radioligand cross-linking and autoradiographic processing
• Immunoreactivity

↝ Prefixation for immunohistochemistry is classically performed *in vivo* by intracardial perfusion of fixative solution (*see* Chapter 7).
↝ Some exceptional antibodies yield positive labeling on unfixed tissue sections; this eventuality should be tested first.
↝ In general, immunolabeling on post-fixed sections is less efficient than on prefixed tissue sections and requires preliminary optimization.
↝ Immunolabeling must first be tested on frozen sections post-fixed with the same mixture as in previous report, using the same antibody with prefixed tissues.
↝ In case of failure, the parameter to optimize is the duration of radiolabeled slide post-fixation.
↝ Test varying balance around half/half concentrations of the two fixatives

6.1.3   Primary Antibodies

Antibodies targeted against tissular antigens are immunoglobulins of the G-type (IgG), which fall into two categories depending on their origin:

• Polyclonal antibodies raised *in vivo* by immunization of an experimental mammal
• Monoclonal antibodies raised *in vitro* by the hybridoma technology, using mouse cell lines

↝ Many of the antibodies used in Western blot do not work in immunohistochemistry — either yielding no specific labeling or recognizing more than one protein.
↝ This information is increasingly, but not always, being specified in commercial catalogs.
↝ To be used at 1:500 to 1:10,000 dilutions

↝ To be used at 1:50 to 1:500 dilutions
↝ Monoclonal antibodies are more specific but less sensitive than polyclonal antibodies.

6.1.4   Signal Amplification

The basic principle of immunohistochemical signal amplification consists of incubating a primary antibody-labeled section with a secondary antibody:
• Raised against immunoglobulins of the animal species used to generate the primary antibody
• Or itself linked with a chromogenic element

↝ The secondary antibody provides amplification because of the existence of two antigen sites per immunoglobulin molecule.
↝ The secondary antibody is generally used at 1:100 to 1:300 dilutions.

6.1.4.1 Peroxidase/Antiperoxidase

Horseradish peroxidase is a vegetable enzyme (extracted from black radish).
After incubation with secondary antibody, slides are incubated with an antiperoxidase antibody that:
• Was generated in the same animal species as the primary antibody
• Is now bound to sections in the ratio of several antiperoxidase molecules per initial antigenic site
Slides are then exposed to peroxidase.

↪ Called "PAP complex"

↪ Small-sized molecule (MW = 40 kDa)

↪ This step is usually fused with the previous one, into a single incubation with a mixture of peroxidase and antiperoxidase of the appropriate species, called the "PAP complex."

Each initial antigenic site now bears a molecular pileup comprising, from tissular antigen to the top of the pile:
• One primary antibody molecule
• Secondary antibody molecules
• Antiperoxidase molecules
• Peroxidase molecules
• Chromogen, if applied thereafter, is then oxidized to colored precipitate.

↪ Classically called the "sandwich technique"
↪ The number of molecules increases from each level to the next, hence amplification.

↪ The most commonly used is diaminobenzidine (DAB).

6.1.4.2 Avidin-Biotin

↪ It is now the most widely used system for immunohistochemical revelation (development).

Avidin is a 68-kDa glycoprotein with very high affinity for the small-molecule vitamin, biotin.
• K_d of avidin/biotin complex is on the order of 10^{-15} M range;

↪ This affinity, still higher than for antigen/antibody, renders the binding of avidin to biotin quasi-irreversible.

• Avidin has four binding sites for biotin.
• Most proteins can be conjugated with biotin, into "biotinylated" forms.
• Corresponding detection system usually involves a biotinylated secondary antibody and an avidin-coupled chromogen.

↪ Allows great amplification of signal
↪ Biotinylated forms of all detection tools are commercially available.
↪ This is the basis of the "ABC kits," nowadays commercialized by several firms. The sections previously labeled with primary antibody first receive the biotinylated antibody and then, after rinsing, the avidin-chromogen construct, both according to manufacturer's recipe.

6.1.5 Chromogens

Whatever the primary antibody, different types of markers are commonly used, including:

↪ Each of these markers is:

- Enzymes
- Fluorescent groups
- Particles

⇨ Either linked to a secondary antibody
⇨ Or bound by a secondary antibody (PAP)
⇨ Or linked to streptavidin, if a biotinylated secondary antibody is used
⇨ All these markers are commercially available for any antibody.

6.1.5.1 Enzymes and chromogen substrates

Enzymes are the most commonly used marker. Enzyme choice can depend on occurrence of endogenous enzymes in tissue.

⇨ Some particularly stable enzymes are preferred and conjugated to all substrates.
⇨ A diversity of chromogens is available for each enzyme marker.

6.1.5.1.1 ALKALINE PHOSPHATASE

This widely used enzyme is extracted from veal intestine.

⇨ Large-sized molecule (MW = 80 kDa)

1. Advantages:
- Numerous chromogen substrates
- Simplicity, reproducibility

⇨ Precipitates with different colors
⇨ Chromogens are available ready-to-use
⇨ Conjugates are available under all forms (IgG, streptavidin, etc.)
⇨ Complementarity with peroxidase

- Multiple labeling

2. Inconveniences:
- Soluble in ethanol

⇨ Avoid dehydration/defatting after immunolabeling.
⇨ Sections can be mounted only with "aqueous mounting medium" (commercially available).
⇨ Colored precipitates lack long-term stability; preparations can be stored briefly at 4°C.

- Cannot be stored for long

- Occurrence of endogenous phosphatase; its activity is blocked by heat (60°C, as for paraffin embedding) or by pretreatment of sections with levamisole.
- Diffusion of colored precipitates around enzymatic sites

⇨ This enzyme is widely present in biological tissues.
⇨ Heat decreases immunogenicity of some tissular antigens.
⇨ Low resolution

3. Chromogen substrates:
- NBT-BCIP

⇨ The two substrates are used simultaneously; both are soluble in dimethylformamide $(CH_3)_2NOCH$.
⇨ $C_{40}H_{30}Cl_2N_{10}O_6$; MW = 817.70
⇨ $C_8H_6NO_4BrClPxC_7H_9N$; MW = 433.60

- NBT (or Nitroblue tetrazolium)
- BCIP (or 5-bromo-4-chloro-3-indolyl phosphate)

⇨ **Caution:** precipitate is soluble in ethanol
⇨ Revelation time can reach 24 h

• Fast-Red	⇒ (5-Chloro-2-methoxy-benzene-diazonium chloride)
	⇒ $C_7H_6N_3O_2$; MW = 250.90
	⇒ Soluble in dimethylformamide $(CH_3)_2NOCH$
	⇒ **Caution:** precipitate is soluble in ethanol
	⇒ Revelation time can reach 24 h

6.1.5.1.2 PEROXIDASE

1. Advantages:

• Numerous chromogens	⇒ Complementarity with alkaline phosphatase
• Multiple labeling possibilities	⇒ Can produce precipitates with diverse colors
• Insoluble in ethanol	⇒ Permanent mounting in resin

2. Disadvantages:

• Chromogens must be prepared extemporaneously	⇒ The classical chromogen for peroxidase is the potent carcinogenic agent DAB, which must be handled with great care.
• Diffuse precipitate	
• Occurrence of endogenous enzyme, which can be inhibited by section pretreatment with peroxidase-inhibiting agent like 2% H_2O_2	⇒ 5 min at room temperature before incubation with primary antibody

3. Chromogen substrates:

• 3´-Diaminobenzidine tetrachloride (DAB) in the presence of H_2O_2	⇒ Highly efficient but very dangerous to handle
– Very intense reaction	⇒ Brown precipitate
	⇒ Can be intensified by reacting in the presence of nickel salts
	⇒ Precipitate not soluble in ethanol
	⇒ Definitive and stable mounting under resin
– Carcinogenic product	⇒ Dangerous wastes; use specific containers.
– Background hazard	⇒ Revelation must be watched (a few minutes are enough) and can be slowed by manipulating on ice.
	⇒ Background can be decreased by pre-revelation incubation in H_2O_2 devoid DAB solution.
– Demands fresh stock H_2O_2	⇒ Stock H_2O_2 must be stored in the dark at 4°C and renewed several months after bottle opening.
• 4-Chloro-1-naphthol	⇒ $C_{10}H_7ClO$; MW = 178.60
	⇒ Blue precipitate, distinguishable from the alkaline phosphatase-reacted NBT-BCIP
	⇒ Soluble in ethanol
	⇒ Poorly stable over time
• 3-Amino-9-ethylcarbazole (AEC)	⇒ $C_{14}H_{14}N_2$; MW = 210.30
	⇒ Red precipitate
	⇒ AEC solvent is toxic by inhalation

Table 6.1 Compilation of Different Chromogenic Enzyme/Substrate Systems

Enzymes	Chromogen Substrates	Precipitates
Alkaline phosphatase	NBT-BCIP	Blue
	Fast-Red	Red
Peroxidase	DAB	Brown
	AEC	Red
	4-Chloro-1-naphthol	Blue-violet

6.1.5.2 Fluorescent markers

Fluorescent molecules possess conjugated double bonds within their structure (*see* Figure 6.1) and react to incident photons by emitting another photon with higher wavelength than the incident photon.

$$-C=C-C=C-$$

Each fluorochrome is characterized by the range of light wavelengths that can activate it (excitation spectrum) and the wavelength of light emitted under activation (emission spectrum).
The most commonly used fluorochromes are:
• Rhodol green
• Fluorescein isothiocyanate (FITC)
• Tetramethyl rhodamine isothiocyanate (TRITC)
• Texas Red sulfonyl chloride
• Aminoethylcoumarin (AMCA)

1. Advantages:
• Rapid procedure

• Easy detection
• Resolution

• Multiple labelings

2. Disadvantages:
• Autofluorescence
• Light absorption
• Photodegradation of fluorochromes

6.1.5.3 Gold particles

Colloidal gold is a commonly used marker for both light and electron microscopes.

↬ This structural property is responsible for the fluorescence under excitation by appropriate light source.

Figure 6.1 Conjugated double bonds characterizing fluorescent molecules.
↬ Filters allow one to separate the emission spectra of the different fluorochromes (wavelengths 300 to 600 nm)

↬ Listed in decreasing order of efficacy

↬ No chemical reaction is required for visualization.
↬ Requires a fluorescence microscope
↬ Best resolutions are obtained at short wavelengths.
↬ Alternating specific filters fitting, respectively, the different fluorochromes

↬ Requires control samples
↬ Weak intensity of images
↬ During observation, duration of which is limited

↬ Gold particles are impervious to electrons.
↬ Gold particles of various sizes are commercially available (1 to 40 nm).

Detection sensitivity decreases with particle size.

➷ It is not well-suited for immunohistochemistry combined with receptor autoradiography because:
  • In emulsion-coated autoradiograms, gold particles are difficult to distinguish from silver grains.
  • In film autoradiograms, the lack of sensitivity and of color will impede interpretation.

6.1.5.4 Choice of marker

Choice is dictated by the following parameters:
• The molecule to be marked

➷ All markers are commercially available for all types of secondary antibodies, but for a subset of primary antibodies

• Endogenous enzyme occurrence in cells or tissue analyzed

➷ Most endogenous enzymatic activities can be inhibited by appropriate section pre-treatment, but it increases the length and complexity of the procedure.

• Section thickness and autoradiographic procedure (with or without post-fixation)

➷ Marker penetration depends on molecular weight.

• Desired sensitivity

➷ Labeling intensity must generally be increased by amplification systems.

• Conservation of labeling

➷ Fluorescent markers are not conserved longer than several days in the dark at 4°C.
➷ Alkaline phosphatase-revealed labelings are conserved in aqueous mounting medium.

• Multiple labelings
• Amplification

➷ Complementary markers
➷ Tissular density of targeted antigen

6.2 STANDARD PROTOCOL

➷ **Sections must never dry out between the beginning and the revelation step.**

6.2.1 Starting Material

Immunohistochemistry coupling to receptor autoradiography, in light microscopy, can involve two categories of tissular preparations:
• Dry, unfixed radiolabeled sections after film autoradiography

➷ Frozen sections from initially unfixed tissue

• Post-fixed, emulsion-dipped sections after photographic processing

➷ Allows full choice for fixatives
➷ Tissue antigens can be altered by prolonged stay of unfixed sections above 0°C.
➷ Glutaraldehyde post-fixation (for radioligand cross-linking) and photographic reagents are likely to hamper subsequent immunolabeling.

6.2.2 Fixation Protocols

6.2.2.1 Buffered paraformaldehyde

The most common efficient fixative is a 4% paraformaldehyde solution in phosphate buffer at pH 7.4.

➷ Paraformaldehyde concentration can be decreased, especially for combination with wet autoradiography, down to 2% for several antibodies classically requiring this fixative.

6.2.2.1.1 PREPARATION

1. Weigh paraformaldehyde	**40 g/L**
2. Add paraformaldehyde to a beaker of distilled water heated to 90°C using a heating magnetic agitator.	**250 mL**

↪ **Under fume hood, because paraformaldehyde vapors are lethal by inhalation.**

3. Add:
- 1 *M* NaOH solution, drop by drop   **40 drops**

↪ Add NaOH until mixture is transparent.
↪ Do not add more NaOH solution than required, so as not to increase fixative pH.

4. In another beaker, put distilled water to dissolve: **250 mL**
- Na_2HPO_4 **12.35 g**
- $NaH_2PO_4 \cdot 2H_2O$ **2.03 g**

↪ Or $Na_2HPO_4 \cdot 12H_2O$, 31.15 g
↪ To make 0.1 *M* Na/Na$_2$ phosphate buffer at pH 7.4, from standard tables.

5. Heat slightly to help dissolve.
6. Mix the two solutions and complete to 1 L with cold distilled water. Filter.

↪ Cool to 4°C before use.
↪ Can be stored several days at 4°C (for light microscopy)

6.2.2.1.2 PROTOCOL

1. Immerse slides in fixative solution.   **At 4°C**   **30 min – 18 h**

↪ Duration of post-fixation is one key parameter to optimize, the other being the chemical composition of fixative (*see* below).

6.2.2.2 Other fixatives

Some efficient fixative chemicals for classical histology, especially glutaraldehyde and picric acid, impair immunohistochemical labeling with various (but not all) antibodies.
- Some antibodies require non-aldehydic fixative, such as:

— Pure acetone	**10 min at –20°C**
— Pure methanol	**5 – 10 min at –20°C**

↪ Some good primary antibodies are rendered totally ineffective by glutaraldehyde above 0.1%.

↪ These protocols are described by some large manufacturers of antibodies for molecular biology and cancer research.
↪ Tissue preservation is better on thicker sections (not less than 20 µm).

6.2.3 Direct Detection Protocol

↪ Rapid protocol: 90 to 120 min

Uses primary antibody directly coupled to marker
1. Rinsing

↪ The following steps are performed at room temperature.

• Buffer: **30 min (at least)**
  – 0.1 *M* phosphate-
  buffered saline (PBS)
  at pH 7.4

↝ Whatever it is, this buffer will be used throughout the protocol (except for revelation using DAB), in addition to the various other reagents.
↝ Alternatively, 0.1 *M* TRIS-buffered saline (TBS) can be used.
↝ Primary antibodies usually display some specific preference among PBS and TBS.

2. Blocking of nonspecific sites
• Buffer supplemented with   **15 – 30 min**
Triton X-100 and with
one and/or other(s) of fol-
lowing:
  – Non-immune serum   **2 – 5%**
  – Skim milk   **5%**
  – Bovine serum albumin   **2 – 5%**
3. Eliminate excess buffer.

↝ Triton X-100 concentration can vary between 0.01 and 0.3%, depending on section thickness (to be tried empirically).

↝ **Inescapable** step for elimination of background labeling. Nonspecific sites are saturated by the added proteic solutions.
↝ Either by pipetting or by delicately wiping the slide with absorbent paper around each section
↝ Sections can advantageously be circled with a hydrophobic product to minimize volume of the incubation drop while avoiding evaporation-facilitating diffusion of medium.

4. Inhibition of endogenous enzymes:
• 10 minutes in appropriate reagent solution

↝ **Optional,** but highly recommended
↝ Depends on the type of marker used

❏ *Levamisol for alkaline phosphatase:*

  – Stock solution 1 *M* in assay buffer
  – Assay solution 10 m*M*

↝ $C_{11}H_{12}N_2S$ (2,3,5,6-tetrahydro-6-phenylimidazole). MW = 240.8
↝ 240 mg/1 mL PBS
↝ To be used extemporaneously

❏ *H_2O_2 2% for peroxidase:*
• To be prepared extemporaneously from commercial 30% stock solution
• Dilute either in assay buffer or in methanol
5. Apply primary antibody:

↝ Must be stored at 4°C in the dark

↝ **Dangerous; wear gloves.**
↝ Complexation of antibody with tissular antigens

• Diluted in PBS or TBS
with 0.2% Triton X-100
and 0.1% proteic additive
either by:
  – Depositing a drop on   **≥80 µL/section**
  horizontal slide
  – Or by immersing
  slides in antibody bath
6. Incubate primary anti-   **2 h at room temp**
body   **or overnight**
    at 4°C

↝ 1:50 to 1:400 for monoclonal antibodies; 1:500 to 1:5000 for polyclonal antibodies
↝ Excess Triton X-100 can generate background or damage sections.
↝ If bath incubation is preferred, use cytomailers as for radioligand binding (*see* Chapter 3).

↝ In humid chamber (closed box including water-dampened paper)

7. Rinsings:
- TRIS-HCl/NaCl   **3 × 10 min**

➥ Good rinsing is necessary to avoid background; tissue sections are not easily rinsed.

8. Further steps:
- Revelation
- Observation

➥ *See* Section 6.2.4, steps 8 to 12.
➥ *See* Chapter 7.

6.2.4 Indirect Detection Protocol

Uses unlabeled primary antibody, to be revealed with additional steps and reagents.

➥ Increases sensitivity

Steps 1 to 7 of Section 6.3 (direct protocol)

8. Apply conjugated or secondary antibody:

➥ Drop incubation in humid chamber

- Diluted 1:100 to 1:300 in   **≥80 µL/section**
  assay buffer without Triton X-100 nor proteic additives
9. Incubate secondary antibody.   **60 – 90 min
                                      room temp**
10. Rinsings:
- In fresh assay buffer   **3 × 10 min**

➥ On horizontal slides or in bath

11. Remove excess buffer.
12. Revelation

➥ Specific protocol for each substrate

❏ NBT–BCIP

- Substrate preparation

➥ Extemporaneous and sheltered from light

- Mix:
- NBT   **0.34 mg
                                  or 4.5 µL**

➥ MW = 817.7
➥ 75 mg/mL dimethylformamide

- BCIP   **0.18 mg
                                  or 3.5 µL**

➥ MW = 433.6
➥ 50 mg/mL dimethylformamide

  – Assay buffer   **qsp 1 mL**
- Deposit NBT-BCIP, 1:250   **100 µL/section**

➥ **Optional**: Add endogenous enzyme blocker levamisole 1 m*M* (i.e., less concentrated than for initial blocking of nonspecific sites).

- Incubate and watch visually until desired color is obtained.   **1 – 24 h, TA
                                  In the dark**

➥ Reaction can be accelerated by placing slides at 37°C

- Stop reaction with distilled water.   **5 min**

➥ If revelation is too weak, deposit another 100-µL drop of NBT-BCIP solution on each section.

- Counterstain if desired

➥ *See* Section 6.1.

- Mount in aqueous medium.

➥ Precipitate is soluble in ethanol.

❏ Fast-Red

- Extemporaneous preparation
- Dissolve:
— Naphthol phosphate   **2mg**
— Dimethylformamide   **200 µL**

– Add 9.8 mL assay buffer (PBS or TBS);
– Add, immediately before use, 10 mg of 1:250 Fast Red; filter.

– Fast-Red solution	**100 µL/section**

- Incubate while watching, so as to stop when required.   **1 – 24 h, room temp, in the dark**
- Rinse in distilled water   **1 min**
- Mount in aqueous medium

↝ **Dangerous compound**

↝ Reaction is finished when red precipitate is clearly visible. Reaction can be accelerated by incubating slides at 37°C.
↝ Stops the reaction
↝ Precipitate is soluble in ethanol

❑ **DAB-peroxidase**
- Preparation of substrate

↝ **Carcinogenic product; wear labcoat, gloves, and mask; clean any trace with household bleach, dip all dishes and instruments in household bleach.**

– DAB	**5 mg**

↝ Diaminobenzidine, $C_{12}H_{14}N_{14} \cdot HCl$ (final concentration 0.05%)

– TRIS-HCl, 50 mM, pH 7.6	**10 mL**

↝ Dissolve 30 min before, on magnetic stirrer, in a beaker wrapped in aluminum foil.

– H_2O_2, 30%	**3 µL**

↝ Add just before incubating slides (final concentration 0.01%)

- **Optional**, to be added before H_2O_2

↝ To increase sensitivity and staining density, *if required*

– $NiCl_2$	**5 mg**

↝ Final concentration 0.01%
↝ Nickel intensification increases background.

• Deposit substrate.	**100 µL/section**

↝ Incubation can also be performed in bath, for decreasing inter-slide variability.

- Watch visually the rise of color.   **1 – 10 min, room temp**

↝ Brown coloration
↝ Too long an incubation produces general brown background.
↝ Coloration can be slowed by incubating on ice.

- Stop reaction by immersing slides in assay buffer.

↝ Counterstaining is possible

- Dehydrate in graded ethanols:

– Ethanol, 70%	**2 min**
– Ethanol, 95%	**3 min**
– Ethanol, 100%	**3 min**

- Transfer to xylene (5 min at least) and mount in resin.

Precipitate is insoluble in ethanol.

❑ **4-Chloro-1-naphthol**
- Substrate preparation

Extemporaneously, sheltered from light

Dissolve:

4-Chloro-1-naphthol	**50 mg**
Ethanol, 95%	**230 µL**

Add:

TRIS-HCl 50 m*M*, pH 7.6   **50 ml**
   – Mix and filter solution;
     then add: H_2O_2 30%   **100 μL**
- Deposit substrate.   **100 μL/section**
- Watch visually the rise of   **5 – 10 min**   ↝ Blue product; check under the microscope
  color.   **at room temp.**
- Stop reaction with assay buffer.   ↝ Counterstaining possible
- Mount in aqueous medium.   ↝ Precipitate is soluble in ethanol

❑ **AEC**   ↝ 3-Amino-9-ethylcarbazole
- Substrate preparation   ↝ Extemporaneously, sheltered from light
   – Dissolve AEC in dime-   **4 mg in**
     thyl formamide.   **1 mL**
   – Add acetate buffer 0.1 *M*,   **14 mL**
     pH 5.2.
   – Filter
   – Add 30% H_2O_2 **150 μL**
- Deposit substrate.   **100 μL/section**
- Watch visually the rise of   **10 min**   ↝ Pink precipitate; check under the micro-
  color.   **darkness**   scope
- Stop reaction with assay buffer.   ↝ Counterstaining possible
- Mount in aqueous medium.   ↝ Precipitate is soluble in ethanol.

6.3   IMMUNOHISTOCHEMISTRY COMBINED WITH WET AUTORADIOGRAPHY

- Dry sections recovered from film autoradiography must generally be post-fixed before immunohistochemistry.
- For wet autoradiography with glutaraldehyde-post-fixed radioligands (*see* Chapter 4), a compromise must be worked out between radioligand cross-linking and antigenicity preservation.

Efficiency of radioligand post-fixation is tested by film autoradiography, by comparing distributions of bound radioligand between two adjacent sections that have been radiolabeled together and then either dried or post-fixed. Proportionality of bound radioligand loss during post-fixation among section areas displaying different labeling densities rules out artifactual redistribution of label.

- If such compromise cannot be found, immunohistochemical labeling can still be performed on sections immediately adjacent to the radiolabeled ones.

↝ Immunohistochemistry can be tried first on unfixed sections, but this succeeds only in exceptional cases.
↝ Glutaraldehyde can still provide complete ligand cross-linking down to 2%.

↝ This test requires a densitometric system of image quantification (*see* Chapter 6).

↝ Sections must not be thicker than 10 μm.

Chapter 7

Analysis of Light Microscopic Autoradiograms

Contents

7.1   ARTIFACTS

7.1.1   Emulsion defects

7.1.1.1   Overall background

This indicates a source of emulsion activation additional to radioactive sections, which can be one of the following:

• Emulsion has been fogged because pack has been inadvertently opened outside of inactinic light.

• Emulsion has been fogged because pack has been stored in the vicinity of radioactive sources.

• Emulsion has been fogged because slide box or autoradiographic cassette was not sufficiently lightproof.

• Emulsion is too old.

↬ Homogeneous cloud of silver grains on, and outside of, slides

↬ It must be kept in mind that nuclear emulsions, especially autoradiographic films, display a significant intrinsic background.

↬ Carefully wrap unused emulsion or films in their original packages, without deleting or damaging any packaging elements, immediately after use in the darkroom.
↬ Store emulsions in a fridge wherein radioactive solutions (either stock or diluted ones) are *never* introduced.
↬ In such a situation, fog is not homogeneous, but more intense on edges of cassette.

↬ Emulsion aging can be slowed down by storage at 4°C; this is specified in notices of liquid emulsions, but not for films. Both types of emulsion can be thus stored for several months.

7.1.1.2 Dots or lines of stacked silver grains

• Lines may have arisen which result from mechanical harm to emulsion face of developed autoradiogram.

↝ Autoradiographic films, after development, yield dots and small lines of intrinsic origin.
↝ Avoid mechanical contact with emulsion face both before and after development, even once dried. This is especially important for films, because histoautoradiograms will be protected by coverslips soon after development.

• Localized thickenings of emulsion can give dotted or aligned silver grains.

↝ Homogeneity of emulsion thickness is a key parameter in autoradiography, which is not perfectly realized even in manufactured surfaces.
↝ Slide drying method after dipping into liquid emulsion has crucial importance.

• Artifactual lines or dots can have arisen from radioactive contamination of cassette inside.

↝ Always protect cassette inside by placing clean paper sheets below and above the slides/film sandwich.

7.1.1.3 Flawed image

• Film was not directly in contact with radiolabeled sections.

↝ For film autoradiograms

↝ Either cassette is defective and has become loose, which can be compensated by thickening cassette content with paper sheets above slides/film sandwich,
↝ Or, the surface made by juxtaposed slides was not completely flat, which can occur when some slide edges accidentally overlay neighbor slide edges, or when slides have different thicknesses. ***Use slides of one single model from one supplier.***
↝ For film autoradiograms

7.1.1.4 Holes within image

• Lack of a zone within image can arise from hand contact with emulsionized face of non-X-ray films. Some of the autoradiographic films are indeed highly fragile during the development step.

↝ Hyperfilm ^{3}H is highly sensitive to mechanical contact, even soft.
↝ During development, the slightest contact will *detach* emulsion from plastic support; if photographic reagents are manipulated in horizontal trays, these films must be developed with emulsionized side facing up.

7.1.2 Section Defects

• Salty deposits on sections after drying can produce such artifacts on film autoradiograms.

↝ This problem is avoided by final rinse in distilled water after post-incubation rinsings. It cannot arise for histoautoradiograms, because such deposits are eliminated through subsequent dehydration/defatting.

- Slide drying before, during or after incubation with radioligand increases nonspecific binding.

↝ Never hold slides out of solutions during binding assay.
↝ This artifact is ruled out by using bath incubation with radioligand (*see* Chapter 3).

7.1.3   Slide Defects

- Increased background can be caused by the following defects in precoating of slides:
  – Gelatin solution too viscous or not fresh

  – Coating performed with contaminated dishes

  – Frozen tissue dust collected together with the sections in cryostat

↝ Always use extemporaneously prepared gelatin solution (unlike the habit in conventional histology labs of storing "ready-to-use" gelatin solution in a fridge for weeks).
↝ Prepare and use gelatin solution in reserved dishes, free of any previous contact with radioactive solutions.
↝ Keep cryostat constantly clean, even during sectioning session.

7.1.4   Binding Defects

- Defect of specific binding can arise from mistakes in radioligand binding conditions.

- Radioligand can be defective.

↝ Some easily forgotten parameters must be accurately respected: pH, temperature and ionic cofactors of assay buffer, and durations of preincubation and rinsings.
↝ Check each new batch for specific binding ability on positive control tissue before experiment *per se*, even with commercially supplied ligands.

7.2   FILM AUTORADIOGRAMS

7.2.1   Methodological Principle

The parameter of autoradiographic images that relates quantitatively to the tissular concentration of radioactivity is the number of silver grains per unit area, which in turn determines the optical density (OD) of the autoradiographic image.

Optical density can be measured indirectly as the proportion of light absorbed by the image relative to a source of known intensity, according to the physical law:

$$OD = -\log I/I_0$$

where I_0 represents the intensity of light incident on the image and I is the light intensity recovered beyond the image.

↝ *See* Chapter 4, Section 4.1.1.

7.2.2 Computer-Assisted Densitometer

Several firms sell computerized systems that allow radioactivity determination from autoradiographic images by:

1. Measuring light intensity transmitted through the autoradiogram placed over a light source:

• Light is measured in each pixel of a CCD video camera (512×512 pixels), as a voltage (V) delivered by a condenser in proportion to the number of afferent photons.

⇢ $V = k \times$ number of photons $\times I$, where I is the current provided to the camera and k is a constant that depends on system characteristics.

• Elementary voltage values are then affected arbitrary values by the computer software, ranging within a linear scale between 0 (minimal light, i.e., dark, no light admission to the camera) and 255 (maximal light).

• Validity of this intensity scaling relies heavily on homogeneity and constancy of light source, and on correct calibration of the system at the beginning of each measurement session, in order to avoid saturation of the CCD camera at low and high intensities.

⇢ Light source is rendered constant in time and space, respectively by an electronic voltage regulator and a diffusing screen. Automatic gain control of the CCD camera must be disconnected.

2. Converting this measure into optical density by reference to the light source intensity

3. Converting the OD into radioactivity by reference to a standard curve established on the images of a series of known radioactivity samples that had been autoradiographed along with the sections to be analyzed

7.2.3 Area Delineation

• Following pixel-by-pixel measurement of afferent light intensity, subfields of the autoradiographic image used to deduce the radioactivity content of the corresponding section area can be delineated over the control screen using the computer mouse.

• Area delineation must obey several general rules.

⇢ This step is highly critical in the presently addressed experimental approach because it can generate dramatic artifacts even from perfect autoradiographic material.

1. Before delineation, anatomical boundaries of the structure of interest must be localized as precisely as possible, by reference to atlases and/or to counterstained section.

↝ Autoradiographic labeling intensity can make difficult the direct superposition of film image and counterstained section. Then section anatomy can be drawn separately, using either a camera or slide projector, and the resulting sketch transferred underneath the corresponding autoradiogram.

Distribution of receptors within anatomically heterogeneous organs (like brain, adrenal capsule, etc.) very often fits with boundaries of anatomical subdivisions. This general observation is merely a corollary to the concept of **target** in the integrative theory of intercellular communications.

↝ Researchers familiar with organ anatomy can usually analyze receptor autoradiograms directly, the above step being reserved for restricted complex areas.

Receptor distribution can be heterogeneous within a single structure.

↝ In such cases, mere anato-histological observation failed to reveal functional subdivisions. This situation is not uncommon, especially in brain.

2. Before delineation, homogeneity of the autoradiographic labeling must be examined carefully within the area anatomically defined above.

↝ Both on the plane of the section and, when several sections are to be compared, between sections.

In case of heterogeneity, test artifact hypothesis by comparing with other sections at the same anatomical level.

↝ Even for qualitative purposes, a receptor autoradiographic assay must always involve several sections of any given structure in the same experiment.

If intra-structure heterogeneity is not artifactual, densitometric quantification of radioactivity must be focused on a subfield wherein labeling intensity is homogeneous.

↝ Otherwise, quantification will average structures with different receptor characteristics, as in assays using homogenates, but with the probability that relative proportions of averaged substructures will vary from one section to another.

There is a minimum threshold to respect in reducing the area wherein optical density is to be measured. "Optical density of image" results are indeed the global light absorption by an assembly of silver grains, the spatial density of which is related to radioactivity concentration. OD must therefore be measured on areas far larger than single-grain size (30 to 50 μm for Hyperfilm ^{3}H, i.e., the finest among film emulsions).

↝ Optical density of the area equalling one single silver grain is that of absolute black; OD of the same surface devoid of silver grains is that of "zero-radioactivity" film background. These measures correspond, respectively, to presence and absence of one radioactive source.
↝ Radioactivity concentration can be deduced from film OD only on the basis of a statistically relevant number of silver grains.

3. Delineate the largest area with homogeneous labeling inside the autoradiographic image of the structure of interest.

As a real example, the method indicated above (method 3) has challenged alternative methods on one single autoradiographic image, generated from a 20-μm-thick section of adult rat olfactory bulb radiolabeled with 0.1 nM dopaminergic D2 ligand ^{125}I-iodosulpride (*see* Figure 6, Examples of Observations).

4. Such measure must be repeated on three to four different but equivalent tissue samples from the same experiment, to yield **one mean value** of tissular radioactivity concentration.

↬ Other criteria for area delineation have been used in the past by various investigators, such as:

• Delineate the histological boundary of the targeted structure on the counterstained radiolabeled section, and apply this area onto the corresponding autoradiographic image. Such an operation is commonly allowed by the morphometric functions usually included in densitometric software (method 2 of Figure 6 in the Examples of Observations chapter).
• Use small, geometrically predefined areas (like circles or squares) repeated as many times as can fit within the anatomical limits of the targeted structure (method 3 of Figure 6 in Examples of Observations chapter).

↬ The presently advised method is called "method 1" in Figure 6 of the Examples of Observations chapter.

7.2.4 Calibration Standards

7.2.4.1 Commercial radioscales

A stuck-together pile of resin blocks with graded radioactive contents is sliced transversely at the same thickness as assayed tissue sections. Each resulting strip is stuck flat onto a clean glass slide affixed to the autoradiographic film beside the slides bearing radiolabeled tissue sections.
Radioactive contents are supplied in radioactivity unit per equivalent tissue protein mass.

↬ Such blocks are commercially available, either presliced or not, for each of the commonly used isotopes: [3]H, ^{125}I, and [14]C.
↬ Make sure to flatten the strip without tearing it.

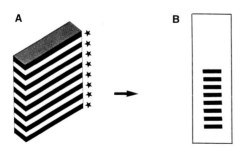

Figure 7.1. A commercial radioscale.
A = Block of piled resin standard bricks.
B = Slide-mounted 20-μm-thick slice of such a block.

7.2.4.2 Homemade standards

Radioactive standards can be prepared in the lab by mixing known, graded quantities of a radioactive compound with given amounts of a tissular homogenate in a series of molds, then snap-freezing these samples and slicing each one in a cryostat.

↪ Homogeneity of the radioactive paste is difficult to obtain, and radioactive contents of the sections used as standards are difficult to determine accurately.

7.3 LIGHT MICROSCOPIC OBSERVATIONS

7.3.1 Principle of the Light Microscope

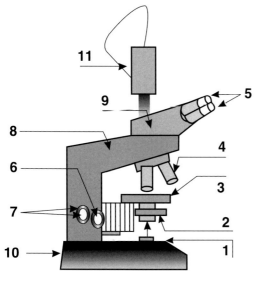

1 = Light source
2 = Condenser
3 = Slide holder
4 = Lens
5 = Ocular
6 = Condenser adjustment
7 = Focus adjustment
8 = Body
9 = Bracket
10 = Base
11 = Image recording system

Figure 7.2 Light microscope.

7.3.2 Bright Field

↪ An ordinary light microscope is called a bright-field microscope.

7.3.2.1 Principle

Light passes through the object before reaching the eye. Microscopic field appears bright, the light of which is partially absorbed by the object. The image observed is more or less dark on a bright background.

↪ Preparation colors depend on the wavelengths selectively absorbed by the object.

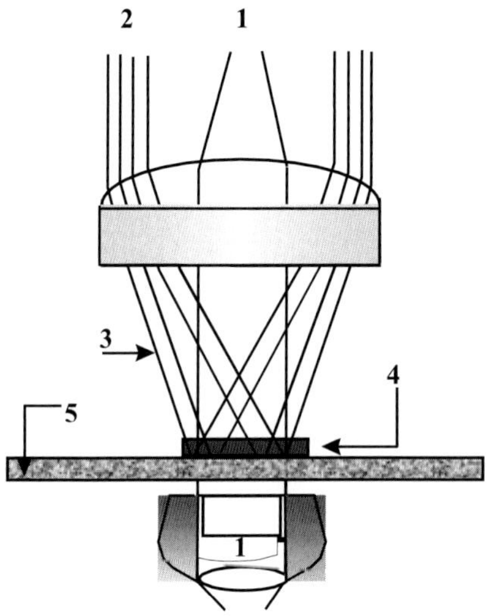

1 = Direct light
2 = Light beams
3 = Diffracted light
4 = Lens
5 = Sample

Figure 7.3 Schematic representation of light beams crossing the preparation (transmission lighting).

7.3.2.2 Utilization

The bright-field microscope is the standard in light microscopy. It allows direct observation of tissular structures.

Silver grains of autoradiographic labeling appear as black dots.

⇨ These small grains can be difficult to detect for beginners, especially in slightly labeled areas.

7.3.3 Dark Field

⇨ Lighting by transmission in dark field

7.3.3.1 Principle

⇨ The difference between dark-field and bright-field microscopes mainly resides in the ratio between openings of lighting and observation.

(1) Dark-field condenser (Schwartzchield): This set of mirrors creates broad and extensive illumination such that the light transmitted directly through the specimen **(2)** is not collected by the lens **(3)**. Only the light that has been diffracted **(4)** by the sample **(5)** is collected by the lens. Dark field is obtained by incident light, flux of which is much lower than in bright field.

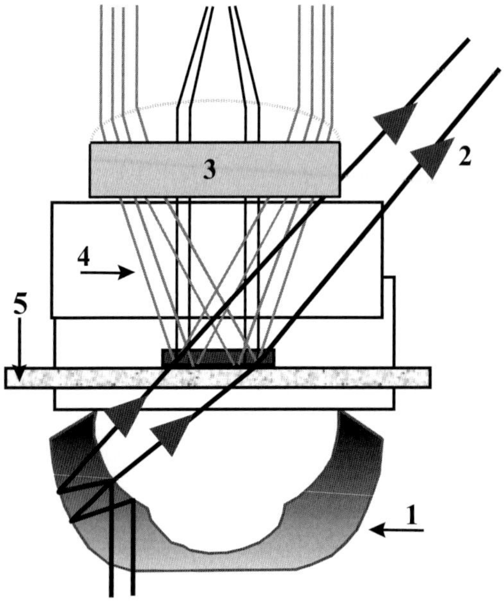

Figure 7.4 Schematic representation of light trajectory through the preparation (dark-field lighting).

7.3.3.2 Utilization

A dark-field microscope is generally used in wet autoradiography to facilitate detection of silver grains generated from fine-grained nuclear emulsions within histological substrata.

↝ Silver grains appear as white dots on a black background.

1. Advantages:
• Usable with low magnification
• Discrimination between silver grains and tissue components
• Radioactivity quantification by silver grain counting

↝ Especially interesting for weak autoradiographic labelings
↝ Either by direct counting or by diffracted light measurement

2. Inconveniences:
• Requires specific equipment
• Requires double observation (dark-field and bright-field)

↝ Dark-field condenser
↝ To localize signal within tissue sample

7.4 QUANTIFICATION OF HISTOAUTORADIOGRAMS

7.4.1 Methodological Principle

The parameter of autoradiographic images that relates quantitatively to the tissular concentration of radioactivity is the number of silver grains per unit area.
In the case of wet autoradiograms, this parameter can be measured by two distinct approaches:
• By optical density measurement, in dark field and at low magnification
• By grain counting, in bright field and at high magnification

↝ *See* Chapter 4, Section 4.1.1.

7.4.2 Dark-Field Densitometry

The computerized systems that were previously described for film autoradiogram analysis also allow radioactivity determination from wet autoradiograms, provided they are observed in dark field. The amount of light refracted into the microscope lens is indeed proportional to the number of silver grains per unit area.
The CCD camera must be adapted to the photographic tube of the microscope.

↝ Only if counterstaining intensity is not high enough to alter light diffraction by the silver grains.
↝ *See* Chapter 4, Section 4.1.1.

7.4.2.1 Measured parameters

1. Measurement of light intensity refracted into the lens by the histoautoradiogram.

- Light is measured in each pixel of a CCD video camera (512×512 pixels), as a voltage (V) delivered by a condenser proportionately to the number of afferent photons.

- Elementary voltage values are then assigned arbitrary values by the computer software, ranging within a linear scale between 0 (minimal light, i.e., dark, no light admission to the camera) and 255 (maximal light).

☞ $V = k \times$ number of photons $\times I$, where I is the current provided to the camera and k is a constant that depends on system characteristics.

- The validity of this intensity scaling relies heavily on the constancy of the microscope light source and on the correct calibration of the system at the beginning of each measurement session.

2. Conversion of refracted light intensity (I) into optical density (OD):

$$OD = \log I$$

☞ Microscope light intensity must be fixed at its maximal value.
☞ Automatic gain control of the CCD camera must be disconnected.

3. Conversion of OD into radioactivity by reference to a standard curve established on the images of a series of known radioactive samples that have been autoradiographed along with the sections to be analyzed.

☞ These standards can be advantageously obtained from commercial resin strips (*see* Chapter 7, Section 7.2.4) by apposition to emulsion-coated coverslips; liquid emulsion slips indeed along resin strips.

7.4.2.2 Area delineation

Following pixel-by-pixel measurement of refracted light, subfields of the autoradiographic image used to deduce the radioactivity content of the corresponding section area can be delineated over the control screen using the computer mouse. Rules for defining areas at which to measure optical density in dark field are identical to those previously listed for film densitometry.

1. *Before delineation,* anatomical boundaries of the structure of interest must be localized, as precisely as possible, by shifting to bright-field examination of the histoautoradiogram.

Receptor distribution can be heterogeneous within a single structure.

☞ In such cases, mere anato-histological observation failed to reveal functional subdivisions. This situation is not uncommon, especially in brain.

2. ***Before delineation*** the homogeneity of the auto-radiographic labeling must be examined carefully within the area anatomically defined above.

In case of heterogeneity, test artifact hypothesis by comparing with other sections at the same anatomical level.

If intra-structure heterogeneity is not artifactual, densitometric quantification of radioactivity must be focused on a subfield wherein labeling intensity is homogeneous.

3. ***Delineate the largest area with homogeneous labeling inside the autoradiographic image of the structure of interest.***

4. ***Such measurements must be repeated on three to four different but equivalent tissue samples*** from the same experiment, to yield ***one mean value*** of tissular radioactivity concentration.

7.4.3   Bright-Field Grain Counting

This method consists of counting the absolute number of grains within one known area. The ratio between these two determinations yields the radioactivity index, to be calibrated by similar assessment of standards (*see above*).

1. ***Grain counting*** can be performed either by manual scoring on paper prints or by computer-assisted stereometry after proper threshold adjustment.

2. ***Area measurement*** is usually provided by morphometry software of image analysis packages.

➥ Both on the plane of section and, when several sections are to be compared, between sections.

➥ Even for qualitative purposes, a receptor autoradiographic assay must always involve several sections of any given structure in the same experiment.

➥ Otherwise, quantification will average structures with different receptor characteristics, as in assays using homogenates, but with the probability that relative proportions of averaged substructures vary from one section to another.

➥ Yields greater resolution than OD assessment

➥ Radioactivity can thus be determined within single cells, or even in subcellular compartments, because of emulsion grain thinness.

Chapter 8

Biochemical
Assay
of Receptors

Contents

8.1 PRINCIPLE

This approach consists of applying the basic protocols for hormone receptor analysis on sections of target organs instead of aliquots of tissue homogenates.

Such transposition enables functional receptor characterization with much higher anatomical resolution than classical approaches using homogenates.

It relies on the abilities to:

• Measure radioactive contents of labeled sections through the optical densities of autoradiographic images

• Make tissue sections that can be assimilated to a series of identical homogenate samples

↬ First introduced around 1980
•*Young, W.S., Kuhar, M.J., 1979, A new method for receptor autoradiography: (^{3}H) opioid receptors in rat brain, Brain Res. 179, 255-265.*
•*Quirion, R., Hammer, R.P., Herkenham, M., Pert, C.B., 1981, Phencyclidine (angel dust)/ μ-opiate receptor: visualization by tritium-sensitive film, Proc. Natl. Acad. Sci. USA, 78, 5881-5885*
↬ Especially adequate for pharmacological studies in central nervous system, it is widely used in neuropharmacology.
↬ Excellent cryostats are commercially available.

8.2 SATURATION ASSAY

8.2.1 Aim

To determine K_d and B_{max} of the receptor in anatomically restricted structure(s).

↬ K_d = Dissociation constant (nM) (*see* Chapter 3, Section 3.2)
↬ B_{max} = Tissular concentration of receptors (*see* Chapter 3, Section 3.2)
↬ *See* Chapter 3 for theoretical background.

8.2.2   Sectioning Protocol

8.2.2.1   Prerequisites

1. Knowledge of the precise biochemical protocol to be run
2. Knowledge of quantitative variations of receptor concentration within the structure to assay, so as the series of sections assayed corresponds to identical samples of reactive tissue

↪ Must be established anew to specifically fit any given assay.

↪ To be determined by a semiquantitative mapping of radioligand binding sites throughout the anatomical structure assayed.

8.2.2.2   Method

1. Select, visually, the appropriate anatomical level of organ on the basis of preliminary receptor mapping (*see* Section 8.2.2.1).
2. Collect the required number of adjacent sections on different slides in prenumbered order.

3. If more than one section per condition is desired, collect a first series of adjacent sections, one per slide (without touch-mounting), and then come back to the first slide and collect a second series of adjacent sections in similar fashion.

↪ So there is no significant variation of receptor concentration within the entire set of sections required for the assay
↪ One series for one saturation assay must include twice as many slides as radioligand concentrations, for total and nonspecific assays.
↪ It is highly recommended to assay at least two sections per condition in any single experiment.

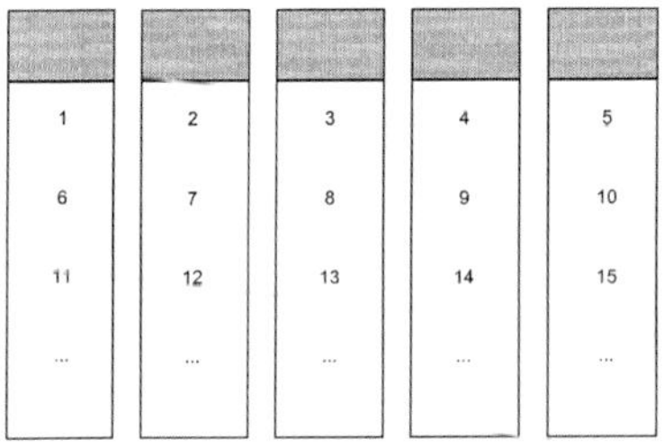

4. Determination of "total" binding for all the radioligand concentrations assayed should be gathered in a median subset of adjacent slides.

Figure 8.1 Serial sectioning protocol.
↪ To ensure that specific labeling parameters will vary as little as possible among the tissue samples to be compared

8.2.3   Radioligand Concentration Range

• Usually $(0.1 \times K_d)$ to $(10 \times K_d)$

↪ On the basis of previous biochemical characterization of the assayed receptor on tissue homogenates

- Eight radioligand concentrations are a good compromise between biochemical habits and autoradiographic constraints.

↪ If the receptor has never been assessed in the tissue of interest, use another receptive tissue as a reference.

8.2.4 Exposure Protocol

1. Expose on the same film sections that have been labeled with radioligand concentrations differing by not more than half an order of magnitude.
2. Slides labeled with a 0.1 to $10 \times K_d$ range of radioligand concentrations will require exposures ranging, respectively, from 20 to 1.
3. Expose on each autoradiographic film of assay a radioactive standard scale, i.e., a series of sections with known and homogeneous radioactive contents.

↪ For quantitative purposes, autoradiograms must be neither saturated (black, i.e., overexposed) nor underexposed.

↪ For absolute determination of radioactivity content of autoradiographed tissue

8.2.5 Data Analysis

1. Quantify bound radioactivity, by densitometry within areas of interest, on autoradiograms from both total and nonspecific slides, according to the principles described previously.

↪ *See* Chapter 7, Section 7.2.3.

Densitometric data

F (nM)	T (mol/mg) *From optical density of a given area of a "total" slide*	NS (mol/mg) *From optical density of the same area of adjacent "non specific" slide*

↪ F = Free (radioligand concentration), calculated from radioactivity measurement on the basis of known "specific activity" of radioligand (*see* Chapter 1)
↪ mg = Milligrams of tissue proteins
↪ NS = Nonspecific radioligand binding
↪ T = Total radioligand binding

2. Subtract nonspecific from total binding at each radioligand concentration. This calculation yields the specific binding B for each radioligand concentration F.

Then apply the Scatchard transformation to specific binding data.

↪ *See* Chapter 2, Section 2.1.2.

Calculations

F (nM)	T – NS = B (mol/mg)	B / F

↪ B = Bound radioligand (to receptors)

3. Plot B/F as a function of B.

↪ Scatchard plot

4. Test it by linear regression: if correlation coefficient is sufficiently close to 1.0, the labeled structure can be concluded to contain one single apparent type of binding site.

The slope of the Scatchard plot is $-1/Kd$.
The abscissa at origin is B_{max}.

↪ K_d = Dissociation constant (same units as F)
↪ B_{max} = Tissular concentration of receptors

8.3   COMPETITION ASSAY

8.3.1   Aim

To determine kinetics of radioligand binding inhibition by a nonradioactive competitor. When several competitors are assessed in parallel, this approach provides the pharmacological profile of the radiolabeled receptor.

↪ *See* Chapter 3, Section 3.1.2 for theory.

8.3.2   Sectioning Protocol

↪ Must be established anew to fit specifically any given assay

8.3.2.1   Prerequisites

Same as for saturation assays.

↪ *See* Section 8.2.2.1.

8.3.2.2   Method

1. Select, visually, the appropriate anatomical level of organ, on the basis of preliminary receptor mapping (*see* Section 8.2.2.1).

↪ So there is no significant variation of receptor concentration within the entire set of sections required for the assay

2. Collect the required number of adjacent sections on different slides in prenumbered order.

↪ One series for one competition assay must include as many slides as competitor concentrations, plus several additional ones for total and nonspecific assays.
↪ It is highly recommended to assay at least two sections per condition in any single experiment.

3. If more than one section per condition is desired, collect a first series of adjacent sections, one per slide (without touch-mounting), and then come back to the first slide and collect a second series of adjacent sections in similar fashion. Defrost all the sections of one slide together, after the last section has been collected.

↪ Unmounted, frozen sections must not be left more than 10 to 15 minutes before touch-mount, to avoid physical alterations.

4. The slides for total binding determinations must be regularly spaced along the whole set. Allow only one or two slides for nonspecific binding determination, at the end(s) of the set.

↪ One "total" among 4 to 5 "inhibited"

8.3.3 Radioligand Concentration

This kind of protocol fits theory best when radioligand concentration equals K_d.

↝ Equal for all slides assessed

↝ That is, at half-saturation of the receptors
↝ In practice, this dose can be advantageously lowered to $0.1\ K_d$.

8.3.4 Competitor Concentration Range

According to theory, competitive inhibition of specific radioligand binding to one type of Michaelian receptor yields a sigmoid curve spreading over 2 orders of magnitude of concentration.

Three concentrations per order of magnitude, regularly spaced along the logarithmic scale, are a good compromise for autoradiographic assays.

↝ *See* Chapter 3, Section 3.1.2.

↝ That is, $1, 2, 4 \times 10^{-n}\ M$

8.3.5 Exposure Protocol

1. Expose a radioactive standard scale on each autoradiographic film of assay.
2. Slides covering the entire range of competitor concentrations must be exposed together, and the exposure time adjusted empirically so that the lower part of the curve is not underexposed too much while total binding is not overexposed.

↝ For absolute determination of radioactivity content of autoradiographed tissue
↝ Expose as long as possible to ensure non-saturated images of "total" slides.

8.3.6 Data Analysis

8.3.6.1 Densitometric quantification

1. Quantify bound radioactivity by densitometry within areas of interest, on autoradiograms from total, drug-inhibited, and nonspecific slides, according to the principles described previously.

↝ *See* Chapter 7, Section 7.2.3.

Densitometric data

[I]	T	T_I	NS
(n*M*)	(mol/mg)	(mol/mg)	(mol/mg)
	From optical density of a given area of a "total" slide	*From optical density of the same area of adjacent "drug-inhibited" slide*	*From optical density of the same area of adjacent "nonspecific" slide*

↝ [I] = Competitor concentration
↝ mg = Milligrams of tissue proteins
↝ NS = Nonspecific binding
↝ T = Total binding in the absence of competitor
↝ T_i = Total binding in the presence of [I] competitor concentration

Then apply the Scatchard transformation to specific binding data.

2. Plot B/F as a function of B.

↝ *See* Chapter 2, Section 2.1.2.

↝ Scatchard plot

8.3.6.2 Direct analysis

Subtract nonspecific from total binding at each competitor concentration. This calculation yields the specific binding B_I for each drug concentration [I].

Calculations		
[I]	$T_I - NS = B_I$	B_I / B_0
(n*M*)	(mol/mg)	

↝ [I] = Competitor concentration
↝ mg = Milligrams of tissue proteins
↝ NS = Nonspecific binding
↝ B_I or T_I = Specific or total binding at I concentration of competitor
↝ B_0 = Specific binding in absence of competitor
↝ Percentages are easier to compare between experiments.

2. Plot specific radioligand binding as a function of competitor concentration.
3. Calculate further the $(B_I/B_0 - B_I)$ ratio, in order to apply the Hill formula, which yields the two key parameters of competition kinetics:

$$\log (B_I/B_0 - B_I) = N \log [I] - N \log IC_{50}$$

↝ IC_{50} = Competitor concentration yielding half-maximal inhibition of specific binding
↝ N = Hill coefficient (*see* Chapter 3, Section 3.1.2)

8.3.6.3 Scatchard analysis

• *Rationale:* Since radioligand "specific activity" is defined as radioactivity amount per molecule of radioligand, the presence of non-radioactive competitor molecules can be considered as decreasing the specific activity of the radioligand. This "theoretical" specific activity can be calculated according to the formula:

$$SA = SA_0 \times F_0 / ([I]+F_0)$$

↝ That is, within the assay medium, the non-radioactive competitor dilutes the radioactive ligand toward tissular receptors.

↝ SA_0 = Real specific activity of the radioligand
↝ SA = Calculated specific activity
↝ F_0 = Real radioligand concentration in assay

↝ [I] = Competitor concentration

Then, each competitor concentration can be transformed into a theoretical "free radioligand concentration" by the formula:

$$F_I = [I] \times SA$$

↝ F_I = "Equivalent" radioligand concentration calculated for [I]

- Subtract nonspecific from total binding at each competitor concentration. This calculation yields the specific binding B_I for each drug concentration [I].

Calculations

[I]	$T_I - NS = B_I$	SA	F_I	B_I/F_I
(n*M*)	(mol/mg)	(Ci/mol)		

- The Scatchard analysis can then be performed as with real saturation data, by plotting B_I/F_I against B_I. This will yield **apparent** K_d and B_{max}.

- Test it by linear regression: if correlation coefficient is sufficiently close to 1.0, the labeled structure can be concluded to contain one single apparent type of binding site.

The slope of the Scatchard plot is $-1/K_d$.

The abscissa at origin is B_{max}.

⇝ SA = Calculated specific activity with [I]

⇝ F_I = Calculated free radioligand concentration

⇝ K_d = Dissociation constant (same units as F)

⇝ B_{max} = Tissular concentration of receptors

⇝ **Limitation:** K_d and B_{max} concern only receptor(s) that are significantly labeled at the nonsaturating radioligand concentration used, i.e., with K_d higher than this concentration.

Caution: The calculated "free radioligand concentrations" have no real existence and are merely a trick to analyze the competition phenomenon. If applied outside the range of [I] covering the sigmoid competition curve, the "saturation" plot will be bell-shaped instead of hyperbolic. Such a "result" would be merely an analysis-derived artifact.

8.4 SEMIQUANTITATIVE APPROACH

8.4.1 Principle

To compare radioactive contents of two distinct sections, or subfields of those, without determining their absolute values

For tissue structures already known to contain one single type of radioligand binding site, this approach allows rapid evaluation of receptor density differences among different tissue samples.

⇝ This is possible if optical densities of the autoradiograms to be compared are both included within the quasi-linear portion of the radioactivity/OD plot.

⇝ Much quicker and less expensive than conventional saturation assay

8.4.2 Sectioning Protocol

In case of heterogeneous organs like brain, anatomical levels of sections must be strictly identical between the samples to be compared.

⇝ This constraint requires preliminary anatomical mapping of the binding sites studied.

8.4.3   Radiolabeling Assay

Using one single radioligand concentration

↪ A nonsaturating one ($<K_d$) to optimize signal-to-noise ratio

8.4.4   Data Analysis

1. Optical density determination of the structure(s) studied
2. Ratio of ODs from the sections to compare

↪ *See* Chapter 7, Section 7.2.

8.5   G-PROTEIN ACTIVATION TEST

8.5.1   Principle

8.5.1.1   G-protein-linked receptor function

G-proteins are intrinsic membrane proteins that can specifically bind the guanine nucleotides GTP and GDP and couple numerous types of ligand-complexed receptors to enzymes generating intracellular "second messengers" (i.e., cyclic AMP, inositol phosphates, diacylglycerol, arachidonic acid).

G-proteins are heterotrimers of α, β, γ subunits, activation of which involves reversible disassembly of α from βγ and exchange of bound GDP for GTP, according to the cycle shown below.

↪ The nature of G-protein recruited by activated receptor is an intrinsic characteristic of this receptor.

↪ Specific GTP/GDP binding to G-proteins obeys the same kinetic laws as receptor/ligand systems and displays similar affinities (nanomolar K_d).

↪ The nature of the α subunit determines the nature of the effector enzyme recruited by the activated G-protein. The α subunit also holds the GTP/GDP binding site.

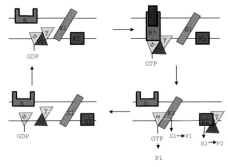

Figure 8.2 G-protein activation cycle.

8.5.1.2 Specific GTP binding significance

Specific GTP binding exclusively characterizes G-proteins that are activated by competent ligand-complexed receptors. Therefore, if a tissue contains one given type of G-protein coupled receptor and if this receptor is "functional" (i.e., able to couple with intracellular effectors), incubation of this tissue with a competent ligand in conditions allowing receptor complexation will display specific GTP binding. Reciprocally, detection of specific GTP binding under agonist exposure will provide an index of G-protein activation and demonstrate the presence of functional receptors in the tissue studied.

↪ First introduced in 1996: *Sim, L.J., Selley, D.E., Dworkin, S.I., Childers, S.R.; Effects of chronic morphine administration on μ-opioid receptor-stimulated ^{35}S-GTPγS autoradiography in rat brain. J. Neurosci, 16(8), 2684-2692.*

↪ Specific binding sites of a radiolabeled extracellular messenger can correspond either to functional receptors (physiological target of the messenger) or to sites of synthesis, degradation, or recycling of the receptors.

8.5.1.3 Autoradiographic application

Radioactive GTP binding on tissue sections and autoradiography allows region-specific detection of receptor-dependent G-protein activation with light microscopic resolution.

↪ The radioligand is γ-^{35}S-GTP.

8.5.2 Protocol

8.5.2.1 Buffers

• *Buffer A*
TRIS-HCl 50 mM, EGTA 0.2 mM, NaCl 100 mM, MgCl$_2$ 3 mM, pH 7.4.
• *Buffer B*
TRIS-HCl 50 mM, pH 7.4

8.5.2.2 Labeling procedure

1. Preincubation in buffer A.   **15 min**
                                      room temp.

2. Charge:
GDP 2 mM + DPCPX   **15 min**
10 μM in buffer A   **room temp.**

↪ DPCPX is an antagonist of adenosine A1 receptor (1,3-dipropyl-8-cyclopentylxanthine).

3. Incubation:
DPCPX 10 μM + GDP   **90 min**
2 mM + DTT 1 mM + ^{35}S-   **room temp.**
GTP 40 pM in buffer A

4. Rinsings:
• Buffer B   **2 × 5 min**
                  4°C

• Distilled water   **Dip**

↪ DTT (dithiothreitol) decreases nonspecific formation of S–S bonds.

5. Air-drying
6. Autoradiography on β-max Hyperfilm

⇝ 3 to 6 days of exposure

8.5.3   Assay

⇝ **Controls**

For any determination of receptor-dependent G-protein activation, three types of incubations with the radioligand must be run in parallel on three subsets of identical samples:
- In the presence of agonist at receptor-saturating concentration
- With radioligand alone
- In the presence of 10 μ*M* non-radioactive GTP

⇝ Receptor-stimulated GTP binding + nonspecific binding
⇝ Basal GTP binding + nonspecific binding
⇝ Nonspecific radioligand binding, to be subtracted from the two above measures

Optional: an additional type of incubation can be added, in the presence of an agonist known to activate G-protein(s) in the tissue sample studied, as a positive control.

⇝ For any part of vertebrate brain, the muscarinic agonist carbachol is a good positive control.

8.6   ADVANTAGES AND INCONVENIENCES

8.6.1   Advantages

- Possibility to determine biochemical parameters on structures below the minimal threshold of macroscopic sampling and homogenizing
- Assessment of anatomical distribution together with biochemical assay in and around the structure studied
- Reliability of the biochemical determinations thus realized
- Possibility to assay simultaneously a large number of structures

⇝ All the more so that radioligand yields low nonspecific binding

8.6.2   Inconveniences

- Greater difficulty and cost than assays using homogenates.
- Lower homogeneity between tissue samples than with homogenates

⇝ In equipment, time, and reagents

⇝ This can be minimized by preliminary distributional studies, judicious sampling, and rigorous data analysis.

Chapter 9

Electron Microscopy

Contents

9.1   PRINCIPLE

The resolution provided by electron microscopy is necessary to determine the specific subcellular components with which the ligand binding protein is associated. Such ultrastructural localization of messenger binding sites is a key parameter to evaluate the functional activity of the messenger in target tissue.

↪ In addition to the subcellular compartment where a given receptor can trigger the physiological response of the cell to its natural ligand, receptors can occur (and be detected) in sites where they cannot couple to intracellular effectors nor even (in the case of hydrophilic messengers) to their ligand:
•Sites of biosynthesis
•Sites of degradation
•Internalization paths
All these sites cannot be resolved from the functional ones at the light microscopic level.

9.2   TISSUE PREFIXATION

Preservation of ultrastructure demands tissue prefixation by the technique presented below, i.e., before animal death — animal is actually killed under anesthesia by fixative perfusion.

9.2.1   Fixative

The classical fixative solution for optimal ultrastructural preservation of mammalian tissues is a buffered aldehydic mixture.

9.2.1.1   Composition

• Paraformaldehyde 4%
• Glutaraldehyde 0.5%
• Sodium/disodium phosphate (Sorensen) buffer 0.1M pH 7.4

9.2.1.2   Preparation (for 1 liter)

1. Weigh paraformaldehyde **40 g/L**
2. Place   paraformaldehyde **250 mL distilled** into a beaker on a heating **water at 90°C** magnetic agitator
3. Add:
• NaOH 1 *M*, drop by drop   **40 drops or until transparency**

⮡ **Under functional fume hood, because paraformaldehyde vapors are lethal by inhalation.**

⮡ Do not add more NaOH solution than required, so as not to increase fixative pH.

4. In another beaker,

⮡ To make 0.05 *M* Na/Na$_2$ phosphate buffer at pH 7.4; from standard tables.

• Distilled water **250 mL**
• Na$_2$HPO$_4$ **12.35 g**
• NaH$_2$PO$_4 \cdot$ 2H$_2$O **2.03 g**

⮡ Or Na$_2$HPO$_4 \cdot$ 12H$_2$O, 31.15 g

5. Heat slightly to help dissolve.
6. Mix the two solutions.
7. Add:
• 25% glutaraldehyde stock solution **20 mL**
8. Bring to 1 L with cold distilled water.
9. Filter.

⮡ Cool to 4°C before use.
⮡ Must be prepared fresh on the day of experiment (for electron microscopy).

9.2.2   Fixation Protocol

9.2.2.1   Perfusion

⮡ Requires sufficient practice to yield satisfactory tissue preservation
⮡ The perfusion must be started as soon as possible after opening the heart to improve tissue preservation.

1. Material:
• Peristaltic pump

• Perfusion cannula
• Mohr forceps
• Regular clamp forceps if dorsal aorta must be clamped
• Usual dissection tools
• Ice-box

2. Anesthetize animal.

3. Adjust flow rate of pump. **150 mL/min**
4. Load it with fixative **at 4°C**

⮡ To allow monitoring of liquid flow rate through plastic tubing between 5 and 200 mL/min
⮡ To fit the diameter of proximal aorta of the animal species used
⮡ To clamp aorta around cannula
⮡ For brain perfusion, to accelerate fixative arrival within nervous tissue
⮡ For fixative solution
⮡ With any of the usual anesthetics (e.g., nembutal, equithesine, chloral hydrate, etc.).

⮡ Fixative solution must be perfused at 4°C to:
⮡ Block endogenous degrading enzymes
⮡ Minimize nonspecific binding in subsequent *in vitro* receptor labeling
⮡ The tubing length must be totally devoid of bubbles.

Keep the switch close at hand.

5. Perfusion

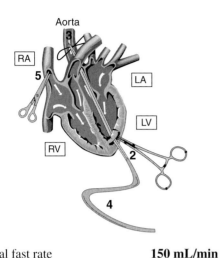

• Initial fast rate   **150 mL/min**
                                           2 min
• Slow rate   **30 mL/min**
                                      7–8 min

6. Immediately after perfusion, remove perfusion tools from animal and proceed.

9.2.2.2   Dissection

❏ *Following steps*
Either cryoprotection

or vibratome slicing

1 = Opening of thoracic cage and pericardium.
2 = Small incision of left ventricle (LV) near heart tip
3 = Introduction of the cannula into the ventricle, delicately up to aorta
4 = Clamping of the cannula with Mohr forceps
5 = Opening of right auricle (RA) with thin, sharp scissors

Figure 9.1 Perfusion technique.
➣ Pump setting at this speed must be checked before starting the experiment
➣ Allow a total perfusion time of 10 min, using 400 mL fixative.
➣ Meanwhile, get out the buffer ice-cubes for subsequent vibratome sectioning (*See* Section 9.4.1).

➣ Immediately after perfusion

➣ Depending on purpose
➣ For preliminary optimization trials (*see* Section 9.3)
➣ For subsequent radioligand binding and processing for electron microscopy (*see* Sections 9.4–9.6)

9.3   PRELIMINARY OPTIMIZATION

9.3.1   Principle

Because receptors are proteins and because tissue prefixation basically involves protein denaturation, technical prerequisites for radioligand binding on tissue sections and for tissular ultrastructure preservation are mutually exclusive. These technical constraints can be solved *a priori* in two ways:

• *In vivo* labeling of receptors by systemic injection of radioligand, and afterward classical tissue treatment for electron microscopy. Such an approach is not detailed in the present book.

⮡ Nonspecific labeling is very difficult to evaluate because it must be assessed on alternate animals receiving an excess of unlabeled competitor along with radioligand. And tissue processing for electron microscopy displays significant inter-animal variability.

⮡ Hydrophilic radioligands cannot reach tissues protected by blood-organ barriers (brain, testis).

• *In vitro* radiolabeling after optimization of a compromise between preservation of tissue ultrastructure and that of binding capacity.

⮡ This method is the only one allowing proper evaluation of nonspecific components among autoradiographic silver grains observed.

9.3.2   Tests

9.3.2.1   Perfusion of different fixatives

The effective fixative is determined by comparing the binding capacities of frozen sections from tissues prefixed by perfusion with various mixtures.
Key parameters include:
• Relative proportions of paraformaldehyde and glutaraldehyde
• Addition of tannic acid, which stabilizes lipids of plasma membranes
• Addition of fixatives selectively stabilizing the tissue studied

⮡ Not above 4% aldehyde(s) altogether

⮡ Improves ultrastructure preservation at aldehyde concentrations lower than 4%
⮡ *See* classical histology textbooks.

9.3.2.2   Cryoprotection and congelation

1. Immersion of prefixed organ in cryoprotective solution:
• Sucrose   **30%**

• Sorensen buffer   **0.1 *M*, pH 7.4**
                                                        4°C
                                                    24 – 48 h

⮡ Cryoprotection takes 24 h for a 1-cm-wide tissue sphere.

2. Afterward, the tissue piece is taken out of cryoprotectant and blotted slightly on absorbent paper.
3. Snap-freezing in isopentane   **–45°C**

⮡ Excess sucrose on tissue will cause a peripheral crust, disturbing cryostat sectioning.

⮡ *See* Chapter 2, Section 2.2.3.3 for snap-freezing

9.3.2.3   Sectioning frozen target organ

Collect series of adjacent sections at anatomical levels containing the receptors investigated.

⮡ For assessment of total and nonspecific binding, in triplicate for one radioligand, on each type of tissue vs. unfixed tissue.

9.3.2.4 Radioligand binding

Total and nonspecific binding must be measured in the same assay using the most pertinent radioligand concentration, i.e., below K_d value down to 0.1 K_d (K_d *is the radioligand concentration yielding half-maximal specific binding: see Chapter 3, Section 3.2*).

↪ This radioligand concentration is far from receptor saturation, in order to minimize nonspecific labeling AND to optimize signal-to-noise ratio. The concentration optimally meeting both criteria must be determined empirically.

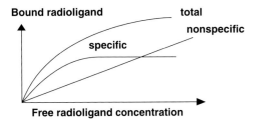

Figure 9.2 Typical saturation kinetics.

Incubation with radioligand and rinsings will be advantageously followed by the post-fixation and dehydration steps required for electron microscopic processing.

9.3.2.5 Specific binding measurement

Either by scintillation counting if structure is large enough among the whole tissue section; or by film autoradiography if receptive structure is too small.

↪ The advantage of using autoradiography is in the opportunity to simultaneously test radioligand specific binding and its retention on several target structures with different binding capacities.

9.3.3 Examples

Prefixation mixture is specifically dependent on the receptor/ligand system. Below are given several prefixatives that have been optimized:

↪ From A. Beaudet's laboratory at the Montreal Neurological Institute (Canada)

- ^{125}I-FK 33824
  - Paraformaldehyde   1%
  - Glutaraldehyde   0.1%
  - Tannic acid   1%
  - Sorensen buffer, pH 7.4   **0.1 M**
- ^{125}I-azido DTLET
  - Glutaraldehyde   0.5%
  - Sorensen buffer, pH 7.4   **0.1 M**
- ^{125}I-DPDYN
  - Glutaraldehyde   0.5%
  - Sorensen buffer, pH 7.4   **0.1 M**

↪ Ligand of μ-opioid receptors

↪ Selective fixative for membrane lipids
↪ Sodium phosphate buffer
↪ Ligand of δ-opioid receptors

↪ Ligand of κ-opioid receptors

- ^{125}I-Neurotensin
  - Paraformaldehyde   **0.75%**   ⟿ Ligand of neurotensin receptors
  - Glutaraldehyde   **0.1%**
  - Tannic acid   **1%**
  - Sorensen buffer, pH 7.4   **0.1 *M***

9.4 TISSUE SAMPLING AND RADIOLIGAND BINDING

9.4.1 Vibratome Slicing

9.4.1.1 Material

A vibratome is a microtome device to slice soft tissue samples. It comprises a fixed object holder and a razor blade holder that is moved at adjustable speed toward the object while vibrating laterally. Both parts are located inside a horizontal bowl (10 cm wide) to be filled with isotonic buffer.

⟿ Slicing with a vibratome requires some practice.
⟿ Ordinary disposable razor blades
⟿ Always use a new blade; clean it with acetone before setting it on the machine, and renew during session.

9.4.1.2 Solution

1. 0.2 *M* stock phosphate buffer
- $NaH_2PO_4 \cdot H_2O$ 0.2 *M*   **27.6 g/L**
- K_2HPO_4, 0.2 *M*   **35 g/L**

⟿ These solutions can be stored at 4°C.

2. Working solution preparation
- Mix 1 volume of $NaH_2PO_4 \cdot H_2O$ (0.2 *M*)
- with 5 volumes of K_2HPO_4 (0.2 *M*) (to ensure pH 7.4).
- Dilute the mixture 1:1.

⟿ The day before experiment. Also prepare buffer ice-cubes by putting in freezer ice-cube trays filled with buffer.

3. Fill the bowl with ice-cold buffer and add several buffer ice-cubes.

⟿ Remove the buffer ice-cubes at the end of animal perfusion.
⟿ Renew ice-cubes (while emptying corresponding liquid volume) regularly during sectioning to keep tissue temperature as low as possible.

9.4.1.3 Tissue holding

Adhere the bottom of tissue piece to clean, dry object holder (glass tray) with a drop of Loctite glue.
After a few seconds, immerse into vibratome bowl.

⟿ Hold tissue piece delicately with blunt forceps.

⟿ Avoid any tissue dehydration.

9.4.1.4   Slicing

Start slicing immediately.
The parameters to adjust are:
• Forward speed
• Amplitude of lateral vibrations
• Object shape and height

⇝ Reshape with scalpel blade, if necessary.

9.4.2   Incubation with Radioligand

Transfer vibratome slices to incubation holes of a chinaware plate, using a flame-shaped glass spatula.

Two incubations must be run in parallel with the same radioligand concentration:
• One in the absence, and
• One in the presence of nonradioactive competitor, yielding total and nonspecific binding, respectively.

⇝ If tissue slice adheres to spatula, detach by shaking gently in medium.

⇝ Prepare one common incubation medium, at the appropriate radioligand concentration. Then split into two identical volumes, one of which will receive 1/100 volume of cold competitor at 10^5-fold the radioligand concentration in order to yield competitor 1000 times more concentrated than radioligand, which ensures thorough inhibition of specific binding (*see* Chapter 3).

9.4.3   Rinsing

Transfer vibratome slices with the glass spatula to a new chinaware plate filled with fresh buffer on ice. Agitate gently with hand.
Then, transfer to post-fixative.
❑ *Following steps*

⇝ (*See* Chapter 5, Section 5.5).

9.5   POST-FIXATION AND EMBEDDING

9.5.1   Radioligand Cross-Linking

9.5.1.1   Principle

For radioligands containing at least one free amine radical ($-NH_2$) within their molecule, radiolabeled vibratome slices are immersed at 4°C for 30 minutes in a phosphate-buffered solution of 4% glutaraldehyde.

For radioligands containing one azido-radical, radiolabeled vibratome slices are exposed 10 min at room temperature under a U.V. light (1 m away from sections).

⇝ Glutaraldehyde cross-linking of bound radioligand onto its specific binding sites yields variable efficiencies among different receptors. Post-fixation efficiency varies from 50% (radioligands of μ-opioid receptors) to 90% (somatostatin receptor radioligand). Efficiency of U.V. cross-linking does not exceed 10%.

⇝ Post-fixation reliability must be checked in preliminary tests.

9.5.1.2   Solutions for glutaraldehyde fixation

1. Sorensen buffer 0.4 *M*, pH 7.4

• $NaH_2PO_4 \cdot H_2O$	**0.718 g**
• $Na_2HPO_4 \cdot 12H_2O$	**12.5 g**
• Distilled water	**100 mL**

2. Preparation of fixative mixture:

• 25% Glutaraldehyde solution	**16 mL**
• Sorensen buffer 0.4 *M*	**12.5 mL**
• Distilled water	**up to 100 mL**

↬ On the day of assay, at least 1 h before use in order to allow fixative to cool.
↬ Heat slightly to help dissolve.

↬ 25% Solution is the commercially available form of glutaraldehyde, to be stored at 4°C.

9.5.2   Rinsing

Transfer vibratome slices with the glass spatula to fresh buffer bath.	**2 × 3 min room temp.**

9.5.3   Tissue Osmication

9.5.3.1   Principle

Additional post-fixation with osmium tetroxide 2%.

↬ OsO_4
↬ **Carcinogen**; commercially available

9.5.3.2   Solutions

1. Buffer
• Dissolve:

– Dextrose	**7 g**
– 0.2 *M* phosphate buffer	**50 mL**

2. Fixative
• Dilute:

– Stock OsO_4 4%	**1 vol**
– Buffer (from above)	**1 vol**

↬ (*See* Section 9.4.1)

↬ **Wear gloves and mask; work under fume hood.**

9.5.3.3   Protocol

Immerse in fixative.	**1 h room temp.**

↬ Place slices flat at the bottom of OsO_4-containing well, since slices will tend to roll up while becoming hard and brittle.

9.5.4   Dehydration

9.5.4.1   Principle

Ultra-thin sectioning demands tissue embedding in hard resins that are hydrophobic, which is possible only with thoroughly dehydrated tissue.

↬ Any water remnant would cause ultrastructural artifacts and compromise tissue preservation.

9.5.4.2   Protocol

Transfer post-fixed slices into small glass vials closed with screw caps, one slice per vial.

Ethanol 50%	**2 × 5 min**
Ethanol 70%	**2 × 5 min**
Ethanol 80%	**5 min**
Ethanol 90%	**5 min**
Ethanol 95%	**10 min**
Ethanol 100%	**2 × 15 min**

9.5.5   Embedding in Epon

9.5.5.1   Principle

To be cut tangentially into semi-thin or ultra-thin sections, vibratome slices are now going to be embedded in a resin that is initially fluid and hardens by polymerization.

↝ Several resins are commercially available, differing mainly in their hardness after polymerization. One example (Epon) is detailed here.

9.5.5.2   Epon preparation

1. Mixture A	
• Epon 812	**31 mL**
• DDSA	**50 mL**

↝ Can be kept for weeks at room temperature in a closed vial

2. Mixture B	
• Epon 812	**25 mL**
• MNA	**22.2 mL**

↝ Can be kept for weeks at room temperature in a closed vial

Final mix:	
Mixture A	**60 mL**
Mixture B	**40 mL**
DMP 30	**1.5 mL**

↝ To be prepared extemporaneously: polymerization begins immediately after addition of DMP 30.

9.5.5.3   Protocol

1. Incubate:	**30 min**
• Epon	**1 vol**
• Propylene oxide	**1 vol**

2. Incubate:	**30 min**
• Epon	**3 vol**
• Propylene oxide	**1 vol**

3. Incubate:	**Overnight**
• Pure Epon	**4°C**

4. Place each slice between plastic coverslips and polymerize 2 h at 60°C.
5. Place the resulting sandwich into Epon-filled plastic cylinder for inclusion.

↝ Classic commercial fabrics for electron microscopy

6. Polymerize:	60°C	↪ Make semi-thin sections and place
• For semi-thin sections	16 – 17 h	blocks at 60° up to a total of 30 h polymerization.
• For ultra-thin sections	30 h	↪ To harden Epon enough for ultra-thin sectioning

9.6   ULTRAMICROTOMY

9.6.1   Preparation of Celloidin-Coated Slides

9.6.1.1   Principle

These slides will be used for reversible mounting of ultra-thin sections on a flat support for metal stainings and dipping into emulsion.

9.6.1.2   Protocol

1. Prepare a 2% solution of celloidin or Parlodion in isoamylacetate.

   ↪ Takes 24 h to dissolve, under constant agitation
   ↪ Make fresh each time.

2. Cross-mark slides with a diamond scribe: 25 mm from one end, 5 mm from each side.

   ↪ These marks will later indicate the position of sections placed on the opposite surface.

3. Immerse clean slides sequentially in:
   • Sulfo-chromic acid   **10 min**
   • Running tap water   **30 min**
   • Distilled water   **30 sec**

   ↪ **Dangerous to handle: wear gloves and lab coat.**

4. Mark the outside of a Borell tube 5 cm from its internal base and filter into it the celloidin solution until the mark is reached.
   • With the scribed end down, dip each slide, drain briefly, allow to dry in a dust-free atmosphere at an angle of 30° (at least 2 to 3 h).

   ↪ With the scribed surface facing downward, using any support about 24 mm high

5. Re-dip the lower end of each slide to a depth of 0.5 to 1 cm and dry as above.

   ↪ This will reinforce the film at the re-dipped end and prevent peeling.

9.6.1.3   Storage

• Celloidin-coated slides can be stored indefinitely in dust-free boxes.

9.6.2   Ultramicrotome Sectioning

9.6.2.1   Principle

Several parameters are involved in section realization:
• Razor orientation

↪ Usually 10 to 15° angle between razor and
section plane

1 = Epon block
2 = Razor
3 = Bending angle

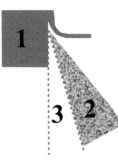

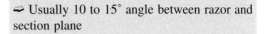

**Figure 9.3 Mutual positions of razor and
block.**

• Sectioning speed

↪ Speed of block movement relative to razor
edge must be determined empirically; it
depends on block shape.
↪ The smaller the section surface, the easier
it is to obtain thin sections.

• Block size and shape

↪ Razor facing side of block ("attack
edge") must be perfectly parallel to razor
edge.

• Razor

↪ Must be kept perfectly clean
↪ All section artifacts caused by razor (i.e.,
scratches, folds, holes, etc.) will affect background labeling.

• Ambient temperature

Block preparation until sections are obtained
involves the following steps:
1. Having fixed the block firmly on block-
holder, trim the block into a trapezium with
the two wide sides perfectly parallel, both
to each other and to razor edge. The widest
side of trapezium is oriented toward the
razor.

↪ Use a new razor blade with single cutting edge.
↪ The two remaining sides allow one to distinguish adjacent sections and to separate
them. Section surface must be reduced as much
as possible (*see* Figure 9.4).

Figure 9.4 Epon block trimming.

2. Direct block toward razor so that inferior block side is parallel to razor edge and block surface is parallel to section plane.

3. Reach area of interest rapidly and then slice until collecting entire section.

4. Fill the razor-delimited receptacle with distilled water, so that the water surface is exactly in the same horizontal plane as the razor edge.

5. Adjust section thickness.

⇒ Orientation is made manually with the aid of razor shadow on the block.

⇒ Particularly with diamond knives, avoid hitting block mass with razor edge.

⇒ Realize reproducibly thick sections, from 0.1 μm (ultra-thin sections, diamond knife) to 1 – 1.5 μm (semi-thin sections, glass knife).
⇒ Ultra-thin section thickness is optimal when sections macroscopically display a silvery-gold color.

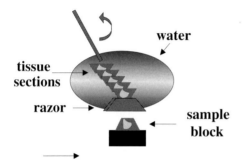

6. Collect ribbons of sections (not isolated sections) on the surface of water by slipping a platinum loop underneath.

Figure 9.5 Making a ribbon of sections.

⇒ Keep the razor clean.
⇒ Like the "special thin loop" n°290 (*Ladd Research Industries, Burlington, Vermont, USA*)

9.6.2.2 Semi-thin sections

1. Trim the Epon block no wider than 3 to 4 mm.

2. Glass razors must have been prepared with a commercial "knife-maker" out of a special glass bar, and resulting razors must be equipped with waterproof topless cisternae.

3. Collect sections on gelatin-coated slides (*see* Chapter 6, Section 6.3)

⇒ 1 μm thickness

⇒ As described in Chapter 8, Section 8.6.2.1.

⇒ Prepare several razors for any single semi-thin section-making session
⇒ Change glass knife as soon as scratches appear on tissue sections.
⇒ *See* Chapter 2 for slide preparation, and Chapter 8, Section 8.6.3 for section mounting on slides.

9.6.2.3 Ultra-thin sections

1. Trim the Epon block no wider than 1 mm.

2. Place diamond knife on ultramicrotome.

⇒ 0.1 μm thickness

⇒ As described in Chapter 8, Section 8.6.2.1.
⇒ Glass knife can also be used, but with more difficulty.

3. Collect sections on celloidin-coated slides (*see* Section 9.6.3).

❑ *Following steps*

↬ *See* Chapter 8, Section 8.6.1 for slide preparation and Chapter 8, Section 8.6.3 for section mounting on slides.

9.6.3 Transferring Sections to Slides

↬ Valid for both semi-thins and ultra-thins

9.6.3.1 Principle

Sections must be allowed to spread on a film of water before adhering to the slide.

9.6.3.2 Protocol

1. Place a drop of distilled water on the horizontal slide.
2. Deposit the ribbon at the surface of the drop.
3. Dry the slide on hotplate.   **40°C**

↬ Just long enough for water drop to evaporate.

For ribbons of ultra-thins, the loop containing the sections is gently lowered onto the opposite surface to one of the inscribed crosses of the celloidin-coated slides.

The edge of the ring is then touched with a small piece of filter paper to draw off the cutting solution, thereby depositing the floating sections on the celloidin.

↬ Deposition is sometimes facilitated by putting a small drop of cutting solution on the slide just prior to placement of the sections.

Repeat this procedure for depositing other sections opposite the other cross.

↬ Care must be taken not to tear the celloidin film over the course of the entire operation.

Store the slides in a dust-free box.

❑ *Following steps*

9.6.4 Staining the Sections

9.6.4.1 Principle

Semi-thins can be stained by classical histological dyes, but electron diffraction under vacuum requires impregnation of ultra-thins by heavy metal atoms.

9.6.4.2 Semi-thin sections
9.6.4.2.1 SOLUTIONS

• Preparation of stain
  – Mix:

↬ For 100 mL

Toluidine blue	**0.5 g**
Sodium borate	**1.0 g**
Distilled water	**100 mL**

- Preparation of sodium maleate buffer   ⇝ For 100 mL

Maleic acid	**0.0116 g**	⇝ To dissolve first in 90 mL distilled water
NaOH 0.5 *N* solution	**A few drops**	⇝ Until pH meter-checked pH rises up to 5.4
Distilled water	**100 mL**	

9.6.4.2.2 PROTOCOL

1. Preincubate in 1% sodium borate.   **2 min**
2. Stain in toluidine blue solution   **3 min**
3. Fresh 1% sodium borate wash   **Dip**

Sodium maleate buffer (*see above*)   **2 × 5 min**

⇝ Add 3 drops of 10% acetic acid per 50 mL in the second bath.

4. Dry.
5. Immerse in xylene.
6. Mount.

⇝ Tissue has already been dehydrated prior to Epon embedding.

9.6.4.3 Ultra-thin sections: double uranyl acetate Reynold's staining

⇝ Ultrastructure must be stained with electron-dense, heavy metal atoms.

9.6.4.3.1 SOLUTIONS

1. Uranyl acetate solution
- Dissolve the day before:
  - Uranyl acetate   **5 g**
  - Distilled water   **100 mL**

⇝ Wrap an Erlenmeyer flask in aluminum foil and place it over a magnetic stirrer.

- Stir overnight in darkened vessel.
- Filter the 5% solution just before use.
- Dilute with 100% ethanol.
  - 5% uranyl acetate   **1 vol**
  - Ethanol 100%   **1 vol**
- Transfer to staining receptacle
2. Stock Reynold's solution
- Dissolve separately:

⇝ For 100 mL
⇝ **Caution: toxic compound!**

  - Lead citrate in distilled water   **2.66 g** / **30 mL**
  - Sodium citrate in distilled water   **2.52 g** / **30 mL**
- Mix the two solutions together and stir 30 min

⇝ Shake the entire contents intermittently.

- Add:
  - Fresh 4% NaOH solution   **16 mL**
  - Distilled water   **24 mL**
- Operating Reynold's stain is diluted:

⇝ Extemporaneously

  - Stock solution   **1 vol**
  - Fresh 0.01 *N* NaOH solution   **1 vol**

9.6.4.3.2 PROTOCOL

1. Uranyl acetate staining protocol:
Place staining mixture in Jolly or Borrel jars.
• Immerse slides so as to avoid mutual contact.

↪ Classical histological dishes with vertical slots can advantageously be used.
↪ Maximum of 15 slides at once

• Stain 20 min in the dark.
• Rinse in 50% ethanol   **6 × 2 min at 4°C**
• Place on a slide rack for drying.
2. Reynold's staining protocol.
• Stain in Jolly jars con-   **3 – 4 min**
taining not more than a
3-cm depth of solution.
• Wash the stain area thoroughly with a stream of distilled water from a washbottle.

↪ Put ethanol baths in ice.
↪ Slides must be thoroughly dried before next step.

↪ With staggered staining times, two slides can be stained per jar.

↪ Do not use the jar and staining solution more than twice.
↪ Test the staining time on grid-mounted sections from the same material to be exposed.
↪ Meticulously avoid the upper end of the celloidin film.

• Allow the slides to dry   **>1 h**
upright in racks, in dust-
free air.
❑ *Following steps*

9.6.5 Carbon Coating

Ultra-thin sections must be coated with a thin layer of carbon, provided by sublimation of carbon from electrodes of an electric arc within a vacuum bell.
The carbon layer should be as thin as possible, which is adjusted by the duration of electric arc operation.
This is controlled by placing in the bell, with the slides, a small piece of filter paper half-covered with vacuum grease. Optimal thickness of carbon layer produces the faintest shade of grey detectable when comparing the two parts of the paper.
❑ *Following steps*

↪ To favor uniform layering of emulsion, to protect it against chemography, to protect stain against photographic reagents, and to stabilize electron beam

↪ Usually seconds to minutes

9.7 AUTORADIOGRAPHY

9.7.1 Dipping into Nuclear Emulsion

9.7.1.1 Material and procedure

↪ Perform under appropriate inactinic light.

As previously described.

↪ *See* Chapter 4, Section 4.3

9.7.1.2　Choice of emulsion

For semi-thins: NTB-2 (Kodak) or K5 (Ilford)　　⤳ Diluted 1:1, as in Chapter 4, Section 4.3

For ultra-thins: L4 (Ilford)　　⤳ Diluted 1:5

9.7.1.3　Reproducibility

For quantitative comparisons between different sections, all samples to compare must be dipped together in the same emulsion batch and with the same handling to ensure identical autoradiographic conditions.

⤳ Estimation of nonspecific labeling proportion on each autoradiogram, required to identify the grains corresponding to specific binding sites through statistical analysis
⤳ Comparison of specific binding capacities between different autoradiograms

9.7.2　Autoradiographic Exposure

⤳ *In absolute darkness at 4°C*

• 4 h (at least) after dipping, store emulsion-coated slides in lightproof slide boxes along with a Dry-Rite bag (one per box); close the box, fix lid with a circular band of black tape, and wrap the sealed box in aluminum foil.
• Split the slides into several sets, to be developed at different exposure times (three sets are ideal; the two longer exposures to be adjusted on the basis of the first one).
• Change Dry-Rite bag monthly in each slide box.

⤳ Avoid any mechanical contact with emulsion coating.
⤳ Dry-Rite bag can be fixed within the box by interposing an additional clean, non-autoradiographed slide.

⤳ Exposure time for ultra-thins is 10-times longer than for cryostat section autoradiograms.

9.7.3　Semi-Thin Autoradiogram Processing

Exactly as wet autoradiograms from frozen cryostat sections.

⤳ *See* Chapter 4, Section 4.3.

9.7.4　Ultra-Thin Autoradiogram Processing

9.7.4.1　Development

⤳ *Under appropriate inactinic light*

9.7.4.1.1　PRINCIPLE

The protocol, as in general photography, is a routine one and must be kept strictly constant. It is never subject to experimental optimization, since its features have already been optimized by emulsion manufacturers.

⤳ Set up water and fixer baths in ice-box 0.5 h before beginning.
⤳ Label the different baths (developer, fixer, water) with permanent marker visible under inactinic light.

9.7.4.1.2 REAGENTS

- Developer (D19 Kodak)
  – Stock D19 solution   **1 vol**
  – Distilled water   **1 vol**

  – Filter diluted solution

↬ When using usual glass staining dishes for histology, each bath requires 200 mL and can accommodate 2 × 20 slides (i.e., 10 slot-rack capacity).

- Fixer (sodium thiosulfate 30%)
  – Sodium thiosulfate   **60 g**
  – Distilled water   **200 mL**
  – Filter diluted solution

↬ Same consideration as for developer bath(s)

9.7.4.1.3 PROTOCOL

1. Developer (D19 Kodak)   **4 min, 17°C**

↬ Cool down from ambient temperature by transferring jar into ice-box; follow up while stirring and remove from ice when thermometer indicates 18°C.

2. Distilled water   **30 sec, 4°C**
3. Fixer   **10 min, 4°C**
   30% sodium thiosulfate
4. Rinse in distilled water **30 min, 4°C**

↬ Can be left in rinsing bath at 4°C, but removal of the preparations should be done the same day to avoid contamination.
↬ Developed slides must not dry out until transfer of ultra-thin autoradiograms onto grids.

5. When ready to remove the preparations from the slides, stand all slides vertically on a clean rack covered by an inverted jar. Place a soaked absorbent paper under the jar to maintain high humidity.

9.7.4.2 Mounting ultra-thins on grids

9.7.4.2.1 MATERIAL

- Round glass container filled to the brim with distilled water, set on a sheet of black paper
- Distilled water
- Glass rod
- Fine forceps
- Mounted needle
- Scissors
- Grids in petri dish with filter paper at bottom
- Razor blade
- Thick filter paper pieces
- Pencil
- Clean petri dishes

↬ Use 150 mesh grids.

9.7.4.2.2 PROTOCOL

1. Lift the thick end of the celloidin film with a razor blade.
2. Scrape both sides of the slide with the blade.
3. Hold the slide against the edge of the glass container in an almost vertical position, with the sections up, and lower the slide into the water.
4. The celloidin film slips from the slide and floats on the surface of the water.
5. Place a grid, using fine forceps, shiny side down, over each group of sections on the floating celloidin film.

6. Write the block reference on a piece of thick filter paper.
7. Breathe on the floating film to hydrate the surface.
8. Curve the filter paper and bring it in contact with the entire celloidin film.
9. Allow the filter paper to become almost completely wet before removing the paper and attached film, with the thinner end of the celloidin being removed first.
10. Cut the excess paper around the celloidin with scissors.
11. Place, paper down, in a petri dish.
12. Place the preparations in a 40°C oven for 1 to 2 hours, partially covered, until completely dried.

9.7.4.3 Thinning the celloidin film on grids

1. Prepare a small glass petri dish, with filter paper at the bottom, half-filled with isoamyl-acetate.
2. Puncture filter-stuck celloidin film around one grid rim with fine-pointed forceps to free the grid from celloidin film.
3. Immerse grid in the isoamylacetate for 120 to 180 sec.
4. Remove and place the section-laden grids in a 40°C oven to dry thoroughly before electron microscopic examination.
❏ *Following steps*
Electron microscopic autoradiograms can now be conveniently stored and treated in the same manner as regular electron microscope sections.

➥ While holding the slide with the section facing downward

➥ The film may be helped in stripping by the use of the mounted needle.
➥ Change water frequently during session and clean by wiping surface with glass rod.

➥ Place grid so that its bars have the same orientation as the sections.
➥ It may help to work under an illuminated magnifier.
➥ One-half of a 10 cm-diameter circle

➥ Apply a continuous, not-too-slow movement so as to avoid folds.

➥ The preparations can be kept indefinitely at this stage if stored in a dust-free atmosphere.

➥ Will promote contrast in the electron microscope

9.8   DATA ANALYSIS

9.8.1   Semi-Thin Autoradiograms

9.8.1.1   Principle

Same principle as for wet autoradiograms from frozen cryostat sections:
- Look for silver grain accumulations within semi-thins, either directly or by initial dark-field observation.
- Ensure whether these accumulated grains correspond to specific binding by checking their absence in the same locus from adjacent vibratome slice radiolabeled in the presence of excess nonradioactive ligand (nonspecific binding).

↝ Cellular resolution is much higher than with cryostat sections.

9.8.1.2   Quantification

- By densitometry in dark-field.
  - Same procedure as for autoradiograms from cryostat sections
- By grain counting in bright-field:
  - Allows one to quantify specific binding capacities in distinct cellular compartments of the tissue

↝ At relatively low magnification
↝ *See* Chapter 7.

↝ At high magnification
↝ Must involve a large enough number of silver grains (i.e., 100 per compartment in any given sample)

9.8.1.3   Contribution to analysis of ultra-thin autoradiograms

Semi-thin quantification allows one to determine nonspecific over total binding ratio in the tissue sample used for electron microscopic analysis.

↝ This parameter is required for the statistical analysis of specific binding subcellular distribution (*see* Section 9.8.2).

9.8.2   Ultra-Thin Autoradiograms

9.8.2.1   Principle

Electron microscopic localization of bound radioligand molecules raises the following problems:
- Single silver grains are wider than some ultrastructural compartments of interest, for example plasma membranes.
- The position of a silver grain cannot be assimilated to the localization of the radioactive source that has activated emulsion.

↝ In the electron microscope, a silver grain appears as a globular bundle of black filaments, with an overall diameter of 200 to 300 nm.

Position of silver grains relative to radioactive source follows a probabilistic law that has been empirically determined for each of the various available isotopes and emulsion/developer couples.

↬ *See Kopriwa, B.M., Levine, G.M., Nadler, N.J., 1984, Assessment of resolution by half distance values for tritium and radioiodine in electron microscopic radioautographs using Ilford L4 emulsion developed by "solution physical" or D-19b methods, Histochemistry, 80(6), 519 – 522.*

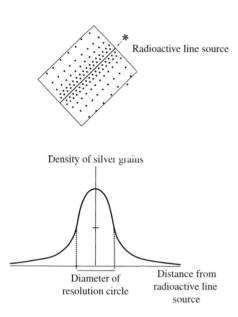

Figure 9.6 Distribution of silver grains around a radioactive line source at electron microscopic level.

Knowledge of this law enables one, reciprocally, to calculate the probability of finding the radioactive source beyond any given distance from the observed silver grain.

↬ For example, with ^{125}I-labeled ligands and with the autoradiographic protocols described above, the probability of finding the radioactive source within the mean diameter of the grain is 50%.

• Silver grains being scattered within tissue ultrastructure, nonspecific labeling distribution must be determined separately in order to reliably subtract background within each subcellular compartment of the tissue sample assayed for total binding.

9.8.2.2   Photography of all silver grains

↬ Compilation of "real grains"

All silver grains present in ultra-thin autoradiographs from "total" and "blank" vibratome slices are systematically photographed at an initial magnification of 10,000 ×.

↬ So as to have a "large number" of grains in each subcellular compartment from each labelling condition, i.e., several hundred grains from total and from nonspecific conditions.

These electron micrographs are then enlarged at the same magnification into paper prints large enough for subsequent analysis (*see* following paragraph).

↬ If one single ultra-thin does not yield a sufficient number of grains, analyze an ultra-thin from another vibratome slice of the same radiolabeling assay.

9.8.2.3   Distribution analysis

- A "resolution circle" of 300 nm diameter, drawn on a transparent overlay, is superimposed over each silver grain.

- The structure ("exclusive grain") or combination of structures ("shared grains") included inside resolution circle are recorded and tabulated.

9.8.2.4   Tissue composition analysis

- A geometrical array of resolution circles, each of which is identical to the one used above for analysis of real grain distribution, is drawn on a transparent overlay that is superimposed on each of the electronographs analyzed above.
- The structure ("exclusive grain") or combination of structures ("shared grains") included inside each resolution circle are recorded and tabulated.

9.8.2.5   Statistical analysis

- The distribution of real grains recorded in sections from slices incubated in the presence of an excess of nonradioactive ligand (nonspecific binding) is first normalized to compensate for differences in tissue composition relative to slices assayed for total binding, according to the normalization equation:

$$Nr_i = r_i \times \frac{H_i \times (\Sigma h_i)}{(\Sigma H_i \times h_i)}$$

Where
Nr_i = Normalized number of nonspecific grains in compartment i
r_i = Number of real grains in compartment i of nonspecifically labeled slice
H_i = Number of hypothetical grains in compartment i of total radiolabeled slice
h_i = Number of hypothetical grains in compartment i of nonspecifically labeled slice
Σ = Sum of all compartments in a given section

➷ Distribution of real grains within the section

➷ This diameter was empirically shown to correspond to 50% probability that the radioactive source was localized within such circle centered on the silver grain.
➷ An "exclusive" grain is a real grain overlaying one subcellular structure only; a "shared" grain is a real grain overlaying at least two subcellular structures.
➷ Distribution of "hypothetical grains" in the tissue section analyzed above.

➷ 12 circles per overlay, so as to obtain about 10 times more hypothetical grains than real grains.

➷ Resulting distribution of "hypothetical grains" yields the relative proportions of the various subcellular compartments in the very tissue samples wherein autoradiographic labeling is observed.

➷ The following calculations are best performed on a computer with appropriate software (not commercially available, to be developed with formulae listed below).

➷ Pioneered by E. Hamel and A. Beaudet (1984): Electron microscopic autoradiographic localization of opioid receptors in rat neostriatum. *Nature*, 312, 155-157.

- The number of nonspecific grains in each compartment is then reduced proportionally to the actual amount of nonspecific binding relative to total binding (established either through radioactivity measurements, or by densitometric analysis of film autoradiograms, or by grain countings on autoradiographed semithins), according to the reduction equation:

$$B_i = Nr_i \times \rho \times \left(\frac{\Sigma \, R_i}{\Sigma \, Nr_i} \right)$$

Where

Nr_i, Σ = As above

B_i = Calculated number of nonspecific grains in compartment i of "total" slice

ρ = Measured nonspecific over total binding ratio

R_i = Number of real grains in compartment i of "total" slice

- The distribution of specific binding in each section incubated with the radioligand alone ("total" slice) is then derived from the specific distribution equation:

$$P_i = \frac{R_i - B_i}{\Sigma \, (R_i - B_i)}$$

Where

P_i = Percentage of specific binding in compartment i

9.8.2.6 Significancy test

Statistical differences among distributions of:
- Real vs. hypothetical grains, in each type of tissue sample (i.e., "blank" and "total")
- Specific vs. nonspecific binding site distributions

are assessed by the Chi-square (χ^2) test.

Chapter 10

Typical Protocols

Contents

10.1 DIFFERENT AUTORADIOGRAPHIC APPLICATIONS FOR THE SAME RADIOLIGAND IN ONE GIVEN TISSUE

Example: labeling of μ-opioid receptors in mammalian brain with ^{125}I-FK-33824.

☛ Hamel, E. and Beaudet, A. (1984) Electron microscopic autoradiographic localization of opioid receptors in rat neostriatum, Nature, 312, 155–157

10.1.1 Overall Comparison

Light Microscopy (film)	Light Microscopy (emulsion)	Electron Microscopy
1. **Snap-freezing** of unfixed brain		1. Light **prefixation** by perfusion of brain with Na_2HPO_4/NaH_2PO_4 buffer at pH 7.4, supplemented with 0.75% paraformaldehyde, 0.1% glutaraldehyde, 1% tannic acid
2. **Cryostat** sectioning		2. **Vibratome** slicing
3. **Incubation** with 0.1 nM ^{125}I-FK-33824 in:		
TRIS-HCl 0.05 M, pH 7.4 1 h at room temperature	TRIS-HCl 0.05 M, pH 7.4, with 0.25 M sucrose 45 min at room temperature	20 min at room temperature
4. **Rinsings** 2 × 5 min at 4°C in:		
TRIS-HCl 0.05 M, pH 7.4	TRIS-HCl 0.05 M, pH 7.4, with 0.25 M, sucrose	
5. **Air-drying**	5. **Post-fixation** In 4% glutaraldehyde in Na_2HPO_4/NaH_2PO_4 buffer at pH 7.4, 30 min at 4°C	5. **Post-fixation** In 4% glutaraldehyde in Na_2HPO_4/NaH_2PO_4 buffer at pH 7.4, 30 min at 4°C. In osmium tetroxide 2%
	6. **Dehydration/defatting** In graded ethanols and xylene	
—	—	7. **Ultra-thin section making and processing**
6. **Autoradiography by affixing** to autoradiographic **film**	8. **Autoradiography by dipping** in nuclear emulsion for light microscopy	8. **Autoradiography by dipping** in nuclear emulsion for electron microscopy

159

10.1.2 Protocols

1. Incubation buffer
- TRIS-base (MW 121.1) **6.03 g**
- HCl 5 *N* **Down to pH 7.4**
- Distilled water **1 L**
- *When necessary*, add: **20 g/L**
  - Sucrose (MW 200.0)

↪ To be prepared extemporaneously
↪ Dissolve in half-final volume water
↪ Add drop-by-drop watching pH meter under agitation, then bring to 1 L with water.
↪ For 0.17 *M* final sucrose concentration

2. Glutaraldehyde fixative mixture
- Sorensen buffer 0.4 *M*, pH 7.4:
  - $NaH_2PO_4 \cdot H_2O$ **0.718 g**
  - $Na_2HPO_4 \cdot 12H_2O$ **12.5 g**
  - Distilled water **100 mL**
- Fixative mixture:
  - 25% Glutaraldehyde solution **16 mL**
  - Sorensen buffer 0.4 *M*, pH 7.4 **12.5 mL**
  - Distilled water **100 mL**

↪ To be prepared extemporaneously

↪ Commercially available solution
↪ Prepared above

3. Osmium tetroxide fixative mixture:
- Buffer
  - Dissolve:
    Dextrose **1 g**
    0.2 *M* phosphate buffer **50 mL**
- Fixative
  - Dilute:
    Stock OsO_4 4% **1 vol**
    Buffer (*see* Chapter 9, Section 9.5.3.2) **1 vol**
- Protocol
  - Immerse in fixative: **1 h** **room temp**

↪ (*See* Chapter 9, Section 9.4.1)

↪ **Wear gloves and mask; work under fume hood.**

↪ Place slices flat at the bottom of OsO_4-containing well, because slices will tend to roll up while becoming hard and brittle.

10.2 BINDING CONDITIONS FOR SELECTED RADIOLIGANDS

All protocols below have been validated for **film autoradiography on cryostat sections.**
Only the specific steps, i.e., preincubation and incubation with radioligand, are described.

↪ For rinsings, autoradiography, and adaptations to other autoradiographic modes, see previous chapters.
↪ Most of the protocols below have been published by Drs. R. Quirion's and/or A. Beaudet's teams (Canada).

Radioligand	Preincubation	Incubation
^{125}I-Neurotensin	None	TRIS-HCl 50 mM, pH 7.4 45 min at room temp.
^{125}I-FK 33824 (μ-*opioid receptors*)	None	TRIS-HCl 50 mM, pH 7.4 1 h at room temp.
^{125}I-D.Ala2-deltorphin-I (δ-*opioid receptors*)	TRIS-HCl 50 mM, pH 7.4, NaCl 150 mM 15 min at 25°C	Same buffer with bovine serum albumin 0.1% 1 h at 25°C
^{125}I-DPDYN (κ-*opioid receptors*)	None	TRIS-HCl 50 mM, pH 7.4 45 min at room temp.
^{125}I-Cholecystokinin	HEPES-NaOH 10 mM, pH 6.5, NaCl 130 mM, KCl 5 mM, MgCl$_2$ 5 mM 15 min at room temp.	Same buffer with EGTA 1 mM, bacitracin 100 μg/mL, soybean trypsin inhibitors 5 μg/mL, phenylmethylsulphonyl fluo- ride (PMSF) 0.1 mM.
^{125}I-Galanin	TRIS-HCl 50 mM, pH 7.4, MgCl$_2$ 5 mM, EGTA 2 mM 30 min at room temp.	Same buffer with bovine serum albumin 1%, leupeptin 0.05%, pepstatin A 0.001% 1 h at room temperature
^{125}I-Neuropeptide Y, ^{3}H-Neuropeptide Y	NaCl 120 mM, KCl 4.7 mM, CaCl$_2$ 2.2 mM, MgSO$_4$ 1.2 mM, KH$_2$PO$_4$ 1.2 mM, NaHCO$_3$ 25 mM, dextrose 5.5 mM, pH 7.4 (Krebs- Ringer phosphate buffer) 60 min at 25°C	Same buffer with bovine serum albumin 0.1%, bacitracin 0.05% 2.5 h at 25°C
^{125}I-Substance P	None	TRIS-HCl 50 mM, pH 7.4, MnCl$_2$ 3 mM, bovine serum albumin 0.02%, bacitracin 40 μg/mL, chymostatin 2 μg/mL, leupeptin 4 μg/mL 1.5 h at 25°C
^{3}H-8-OH-DPAT (5HT1A serotonin receptors)	None	TRIS-HCl 0.17 M, pH 7.7, CaCl$_2$ 4 mM 1 h at room temp.
^{3}H-5HT (5HT1B serotonin receptors)	None	TRIS-HCl 0.17 M, pH 7.7, ascorbic acid 0.1%, CaCl$_2$ 4 mM, fluoxetin 10 μM, pargylin 10 μM 1 h at room temp.
^{3}H-Ketanserin (5HT2 serotonin receptors)	None	TRIS-HCl 50 mM, pH 7.7 1 h at room temp.

Radioligand	Preincubation	Incubation
[3]H-Methyl-carbamyl-choline (nicotinic receptors)	None	TRIS-HCl 50 mM, pH 7.4, NaCl 120 mM, KCl 5 mM, CaCl$_2$ 2 mM, MgCl$_2$ 1 mM 1 h at room temp.
[3]H-Cytosine (nicotinic receptors)	None	TRIS-HCl 50 mM, pH 7.4 1 h at room temp.
[3]H-Pirenzepine (M1 muscarinic receptors)	NaCl 120 mM, MgSO$_4$ 1.2 mM, KH$_2$PO$_4$ 1.2 mM, glucose 5.6 mM, NaHCO$_3$ 25 mM, CaCl$_2$ 2.5 mM, KCl 4.7 mM, pH 7.4 (Krebs buffer) 15 min at room temp.	Same buffer 1 h at room temp.
[3]H-Acetylcholine (extemporaneous ^{3}H-choline acetylation) (M2 muscarinic receptors)	NaCl 120 mM, MgSO$_4$ 1.2 mM, KH$_2$PO$_4$ 1.2 mM, glucose 5.6 mM, NaHCO$_3$ 25 mM, CaCl$_2$ 2.5 mM, KCl 4.7 mM, pH 7.4 (Krebs buffer) 15 min at room temp.	Same buffer 1 h at room temp.
[3]H-AF-DX-384 (M2 muscarinic receptors)	NaCl 120 mM, MgSO$_4$ 1.2 mM, KH$_2$PO$_4$ 1.2 mM, glucose 5.6 mM, NaHCO$_3$ 25 mM, CaCl$_2$ 2.5 mM, KCl 4.7 mM, pH 7.4 (Krebs buffer) 15 min at room temp.	Same buffer 1 h at room temp.
[3]H-Hemicholinium (choline uptake sites)	None	TRIS-HCl 50 mM, NaCl 300 mM, pH 7.4 1 h at 4°C
[3]H-AH5183 (acetylcholine vesicular transport site)	TRIS-HCl 50 mM, pH 7.4, NaCl 120 mM, KCl 5 mM, CaCl$_2$ 2 mM, MgCl$_2$ 1 mM 30 min at room temp.	Same buffer 1 h at room temp.

Examples
of Observations

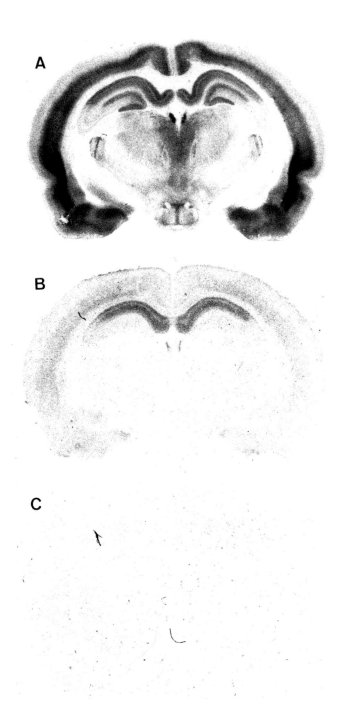

Figure 1
Film autoradiograms of adjacent, transverse sections of rat forebrain labeled with ^{125}I-Tyr^0-D-Trp^8-somatostatin-14 alone (A, total binding); or in the presence of micromolar sandostatin (B, SST1/SST4 binding); or in the presence of micromolar somatostatin-14 (C, nonspecific binding). (*Photographs courtesy of S. Krantic and R. Quirion.*)

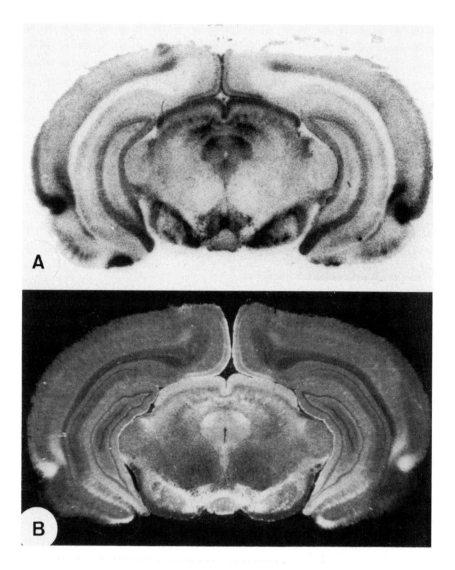

Figure 2

Light microscopic autoradiograms of ^{125}I-neurotensin-labeled, adjacent, transverse sections of rat midbrain. (A) Film autoradiogram in bright field; (B) Emulsion-dipped histoautoradiogram in dark field. (From Moyse, E., Rostene, W., Vial, M. et al., Distribution of neurotensin binding sites, *Neuroscience*, 22(2), 525-536, 1987. With permission.)

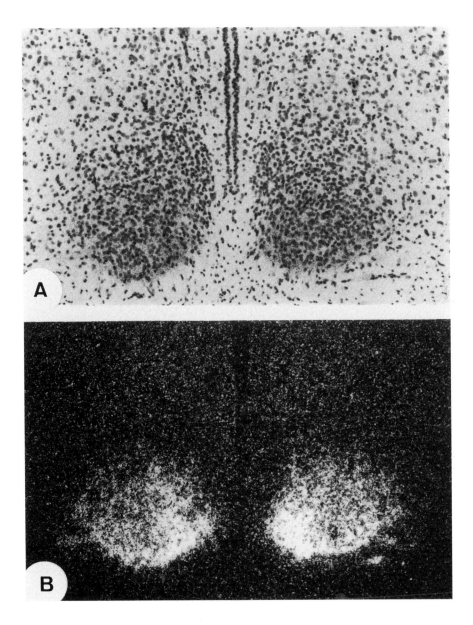

Figure 3
Bright-field (A) and dark-field (B) examination of the suprachiasmatic nucleus in one emulsion-dipped histoautoradiogram from a ^{125}I-neurotensin-labeled transverse section of rat forebrain. (From Moyse, E., Rostene, W., Vial, M. et al., Distribution of neurotensin binding sites, *Neuroscience*, 22(2), 525-536, 1987. With permission.)

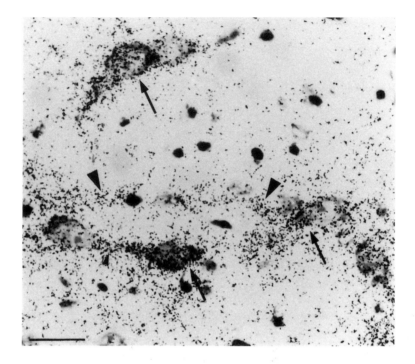

Figure 4

Light microscopic autoradiogram of transverse section of rat forebrain labeled with ^{125}I-neuro-tensin. After incubation with the radioligand, the sections were fixed with 4% glutaraldehyde, defatted, and autoradiographed according to conventional dipping techniques. Intensely labeled neuronal perikarya (arrows) are detected over low background neuropil reaction. Note the presence of clearly outlined labeled processes (arrowheads) in this area, which corresponds to substantia innominata (nucleus basalis magnocellularis). Scale bar: 30 μm. (*Photograph courtesy of A. Beaudet and E. Szigethy.*)

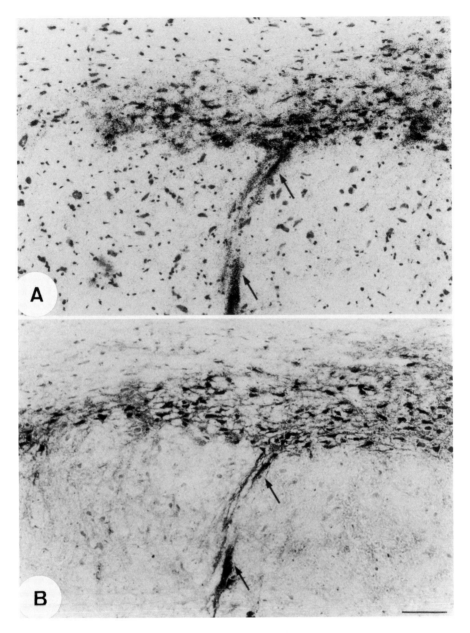

Figure 5

Combined autoradiographic localization of ^{125}I-neurotensin-labeled binding sites and immunohistochemical detection of tyrosine hydroxylase immunoreactivity in adjacent 5-μm-thick sections from the rat substantia nigra (midbrain). Tyrosine hydroxylase (TH) is the biosynthetic enzyme of the neuromediator dopamine, and therefore a specific marker for dopamine-secreting neurons. (A) In the autoradiogram, specifically labeled neurotensin binding sites are detected over the perikarya and dendritic processes of neurons of the pars compacta, as well as over a few isolated cells in the pars reticulata (arrows). (B) ^{125}I-neurotensin-labeled cells are identified as being TH-immunoreactive in the adjacent, immunohistochemically processed section. Scale bar: 100 μm. (*Photographs courtesy of A. Beaudet.*)

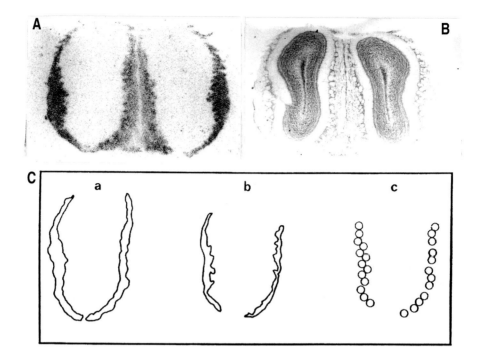

Figure 6

Quantification of D2-like radioligand binding by three distinct densitometric methods. (A) Film autoradiogram generated from section (B). (B) Transverse section of frozen rat olfactory bulb counterstained with cresyl violet after *in vitro* labeling with 0.1 nM ^{125}I-iodosulpride.(C) Three modes of area delineation for autoradiographic labeling quantification by computer-assisted densitometry (see Chapter 7, Section 7.2.3). (From Moyse, E. and Coronas, V., *In vitro* localization and the pharmacological characterization of receptors by radioligand binding, CRC Press, Boca Raton, FL, 1997. With permission.)

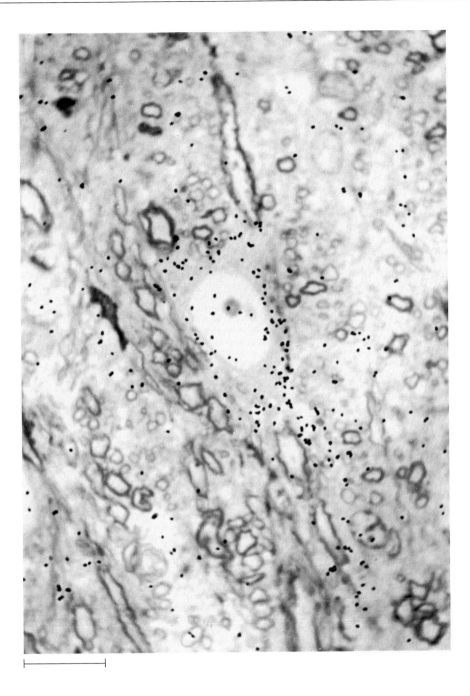

Figure 7
Light microscopic autoradiogram of a 1-μm-thick Epon-embedded section cut from the surface of a midbrain slice incubated with ^{125}I-neurotensin. Some silver grains clearly outline the peripheral delimitation of one neuronal perikaryon (arrow), while others are scattered throughout the neuropil. Scale bar: 10 μm. *(Photograph courtesy of A. Beaudet.)*

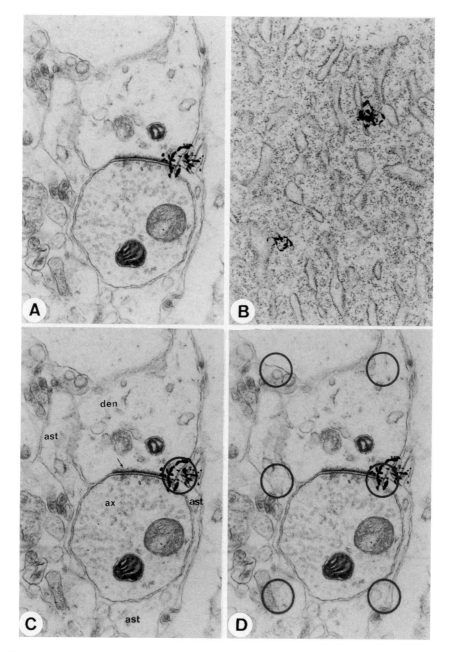

Figure 8

Electron microscopic autoradiograms of sections from rat locus coeruleus labeled with the morphinic agonist ^{125}I-FK-33824 (total binding). Initial magnification 8000×. (A) One "shared" grain associated with an axodendritic asymmetrical synapse. (B) One "exclusive" grain associated with endoplasmic reticulum of a nerve perikaryon. (C) Resolution circle analysis of an observed ("real") grain (D) Resolution circle analysis of tissue composition using an array of "hypothetical grains" superimposed over each of the fields displaying real grains. (From Moyse, E., Marcel, D., Leonard, K., and Beaudet, A., Electron microscopic distribution of mu opioids, *Eur. J. Neurosci.*, 9(1), 128-139, 1997. With permission.)

Glossary

A

Acid
⮑ Chemical substance that spontaneously generates H^+ ions in aqueous solution.

Affinity
⮑ Quantitative measure of the ability of one chemical to form a complex with another.

Allostery
⮑ Reversible alteration of functional properties of a multimeric protein resulting from modification of sterical relationships between the subunits.

Amino acids
⮑ Molecules containing a free carboxyl (COOH) and a free unsubstituted amino (NH_2) group on the α-carbon atom.

Antibody
⮑ Protein that belongs to the immunoglobulin family, and is secreted by mammalian lymphocytes, and that binds with high affinity and specificity to one protein (called antigen).

Association
⮑ Chemical reaction resulting in the formation of a complex.

Astrocyte
⮑ *See* glial cells.

Autoradiogram
⮑ Image produced through autoradiography.

Autoradiography
⮑ Detection of radioactive molecules within a sample by apposition of a photographic emulsion.

Axon
⮑ Subcellular compartment of a neuron that is defined structurally by (1) small diameter (0.5 to 1 μm), (2) presence of specific cytoskeleton elements, (3) absence of ribosomes and perikaryal organelles except for mitochondria, and (4) presence of the voltage-gated channels of nervous action potential.

B

Base
⮑ Chemical substance able to combine covalently with H^+ ions and thus neutralize acids.

Binding site
⮑ Tissular chemical able to form a complex with a ligand; corresponds to a receptor only if binding kinetics can be correlated with triggering of a biological response.

Biogenic amines
⮑ Biologically active molecules containing the amine (NH_2) function.

B_{max}
⮑ Maximal binding capacity; reflects total number of binding sites.

Bond
⮑ Physical link between atoms within a molecule or between subunits within a multimeric protein or between molecules within a complex.

Buffer	↪ Substance or combination of chemicals that keeps aqueous solutions at a constant pH throughout a wide range of acid/base additive concentrations.

C

CCD camera	↪ One among the several standard types of video cameras that provides high resolution (512×512 pixels) and quantitative index of light intensity in each pixel.
Chemical equilibrium	↪ Reversible transformation of chemicals called reagents into other chemicals called products, yielding stable coexistence of both reagents and products.
Chromatography	↪ Physical separation of molecules within a fluid mix by percolation, driven either by gravity, pressure, or electric field, through the pores of an inert substratum (gel or filter paper).
Chromogen	↪ Substance that can be transformed into a colored precipitate, for histochemistry.
Competition	↪ Binding assay purported to characterize dose dependence of radioligand binding inhibition by a steric analog.
Competitive inhibition	↪ Inhibition of ligand binding by a chemical that shares steric similarity with the ligand, thus conferring on it affinity for the same receptor site.
Competitor	↪ Chemical inhibiting ligand binding by competition.
Complex	↪ Pile-up of distinct molecules bound together without chemical modification.
Complexation	↪ Reversible mutual binding of distinct molecules into a complex.
Conjugation	↪ Covalent fusing of two molecules into a single one.
Contrast	↪ Ratio between areas of different optical densities within a photograph or an autoradiogram.
Counterstaining	↪ Histological staining of a preparation after autoradiography.
Covalent bond	↪ Strong inter-atomic bond, breaking of which causes destruction of corresponding chemical substance.
Coverslip	↪ Thin glass lamella stuck to histological preparation for optimizing light beam transmission and for protecting the tissue sample.
Cryostat	↪ Freezing microtome.
Cytokine	↪ Chemical messenger secreted by lymphocytes, but also by virtually any type of cell with molecular mass of 20 kDa or less.

D

Darkroom ⤳ Room devoid of daylight; used for photographic treatments.

Dendrite ⤳ Subcellular compartment of neurons consisting of cell body expansions, containing the same inclusions as cell body and varying in diameter.

Developer ⤳ Chemical that reduces photoactivated Ag^+ ions of photographic emulsion (i.e., latent image) into metallic silver grains.

Development ⤳ Action of developer on photographic emulsion; revelation of latent image.

Dipping ⤳ Coating of a radiolabeled tissue slice with liquid emulsion, for high-resolution autoradiography.

Dissociation ⤳ Chemical reaction liberating previously complexed molecules.

Dissociation constant: K_d ⤳ Radioligand concentration necessary to saturate half of the total number of binding sites in a given preparation.

Dry autoradiography ⤳ Autoradiography by apposition with dry, ready-to-use films.

E

Electronography ⤳ Photography with an electron microscope.

Electrophilic substitution ⤳ Chemical reaction involving the replacement of an atom or a radical by another one with higher affinity for peripheral electrons of the neighboring atoms.

Embryonic inducer ⤳ Chemical messenger contributing to determination of the future fate of embryonic cells.

Emulsion ⤳ Hydrophilic mixture of photoactivatable molecules (usually halogenated salts), for photography or for autoradiography.

Equilibrium ⤳ A reversible chemical reaction that allows stable coexistence of reagents and products.

F

Film ⤳ Transparent (plastic) substratum covered with dried light-sensitive emulsion.

Fixation *(photography)* ⤳ Step of revelation that allows neutralization of developer and dissolution of nonreacted emulsion.

Fixation *(histology)*   ↪ Chemical treatment of living samples for preservation and microscopic observation of cellular structures.

Fixative   ↪ Buffered, isotonic chemical mixture for histological fixation.

Fixer   ↪ Strongly oxidant chemical mixture for photographic fixation.

G

G-protein   ↪ GTP-binding protein.

Gelatin   ↪ Animal-derived protein with high molecular mass endowed with sticky properties.

Gelatination   ↪ Slide coating with gelatin/texture agent (chromium-potassium sulfate) solution for light microscopic histology.

Glial cells   ↪ Nonexcitable cells of nerve tissues (astrocytes, oligodendrocytes, and microglial cells in vertebrate brain).

Growth factor   ↪ Chemical intercellular messenger modulating cell proliferation and/or survival and/or differentiation.

H

HPLC   ↪ High performance liquid chromatography.

Histoautoradiogram   ↪ Radiolabeled tissular section that has been autoradiographed by dipping in liquid emulsion.

Histochemistry   ↪ Visualization of an intrinsic molecule through a coloration resulting from its chemical activity toward an exogenous substrate.

Histology   ↪ Preparation of biological samples for microscopic examination.

Hormone   ↪ Chemical intercellular messenger secreted and diffusing freely into interior medium of organism.

I

IC_{50} ↪ Drug concentration necessary to inhibit 50% of considered maximal response.

Immunocytochemistry   ↪ Visualization of an endogenous chemical by antibody labeling and subsequent staining, on cell cultures.

Immunohistochemistry   ↪ Visualization of an endogenous chemical by antibody labeling and subsequent staining, on tissue sections.

Iodination

⤳ Ligand labeling by addition of radioactive iodine isotope.

Ionic strength

⤳ Ion concentration in an aqueous solution.

K

K_d

⤳ Dissociation constant of a chemical equilibrium.

K_i

⤳ Receptor binding inhibition constant; reflects K_d.

Kinetics

⤳ Thermodynamic laws of chemical reactions.

L

Latent image

⤳ The subset of activated crystals in a photo-sensitive emulsion before developer action; will be converted into silver grains through development.

Law of mass action

⤳ Thermodynamic law stating that in an equilibrium, the concentrations of the products are inversely related to the concentrations of the reagents by a constant called the "dissociation constant" or K_d.

Ligand

⤳ Chemical able to bind specifically the recognition site of receptors, enzymes, etc.

M

Micrography
Microtome

⤳ Photography through the light microscope.
⤳ Machine allowing thin slicing of tissue samples for histology.

N

Nervous system

⤳ Communication system of animals allowing topographically channeled and rapid flow of information within organisms by propagation of elementary bioelectric signals (action potentials) along interconnected excitable cells (neurons).

Nervous tissue
Neuron

⤳ The tissue constituting all nervous systems.
⤳ Excitable cell with long, thin, and stable expansions that (1) propagates unitary depolarizations (nervous action potentials) along its membranes, (2) secretes chemical messengers (neurotransmitters) as a function of depolarization, and (3) is capable of organizing with interconnected nets of communication.

Neurotransmitter	⤳ Chemical intercellular messenger secreted by a neuron.
n_H: Hill constant	⤳ Thermodynamics-derived constant that reflects the number of different binding sites for a given ligand.
Nonspecific binding	⤳ Irreversible and unsaturable component of radioligand binding.
Nucleosides	⤳ Molecules containing a nitrogenous heterocyclic base and a sugar (pentose).
Nucleotides	⤳ Nucleosides containing an additional molecule of phosphoric acid.

O

O.C.T.	⤳ Generic name for a commercial material that is viscous at room temperature and reversibly solidifies below 0°C, and that is used as a "mounting medium" to adhere frozen samples to the object holder of the cryostats.
Optical density	⤳ Quantitative, unitless index of light absorption by a sample (liquid or solid).
Oxidation	⤳ Chemical reaction consisting of electron loss.

P

Paraffin embedding	⤳ Inclusion of an inert, dehydrated sample within a block of paraffin, by embedding the sample in liquid paraffin (at 65°C) within a mold and letting it cool down for further microtomy; the basal process for light microscopic histology.
Peptides and proteins	⤳ Molecules built from amino acids.
Perfusion	⤳ *In vivo* infusion of a exogenous liquid into general circulatory system of a living, anesthetized animal that is usually killed by the process itself.
Pharmacological profile	⤳ Rank order of drug efficiency on either biological responses or receptor binding.
Pharmacology	⤳ Branch of biological sciences devoted to the mode of action of drugs (endogenous as well as exogenous).
Photography	⤳ Image-making using a light-sensitive emulsion.
Post-fixation	⤳ Fixation of histological sections from fresh-frozen samples.
Prefixation	⤳ Fixation of tissue before section-making, usually by perfusion of fixative in living, anesthetized animal.

R

Radioactive isotope	➷ Unstable atomic form of some chemical elements that spontaneously disintegrates with emission of electromagnetic radiation.
Radioactive isotope emission energy	➷ Energy produced during radiation.
Radioactive isotope half-life	➷ Period over which half of the initially present radioactive atoms have returned to their stable chemical form.
Radioactive isotope specific activity	➷ Amount of radioactivity per mass unit.
Radioactive source	➷ Molecule containing radioactive isotope.
Radioactivity	➷ Ability of a substance to emit radioactive radiation.
Radioligand	➷ Ligand labeled with a radioactive isotope.
Reaction	➷ Phenomenon resulting in physical disappearance of some chemicals (called reagents) and concomitant appearance of other chemicals (called products).
Receptor	➷ Molecule that specifically binds a chemical messenger and consequently triggers a biological response of the cell.
Resolution	➷ Smallest distance for distinguishing two separate points under a microscope.
Revelation	➷ *See* Development.

S

Saturation	➷ Binding assay set up to characterize the affinity (K_d) and maximal binding capacity (B_{max}) of the radioligand.
Scatchard analysis	➷ Determination of K_d and B_{max} by measuring bound radioligand for various radioligand concentrations.
Scintillation counter	➷ Machine for radioactivity quantification that converts the frequency of radiation-induced scintillations on a sensitive screen into absolute quantity of radioactivity.
Second messenger	➷ Intracellular molecules produced under the influence of extracellular hydrophilic ligands (first messengers).
Sensitivity	➷ Ability of a photographic emulsion to be activated by weak light.
Silver grain	➷ Metallic silver grain produced by developer in an activated emulsion.
Snap-freezing	➷ Freezing by sudden dipping into a low-temperature fluid.
Sorensen buffer	➷ The monosodium/disodium phosphate buffer most commonly used in histology.

Specific binding	↪ Amount of total binding measured with a given radioligand corrected for nonspecific binding.
Specificity	↪ Binding exclusivity of a molecule toward its ligand(s).
Standard	↪ Sample with known (radioactivity) content for calibration, allowing one to convert a unit-less index into absolute quantities.
Steroids	↪ Phenanthrene derivatives.
Synapse	↪ Morphologically differentiated intercellular junctions of neurons with other neurons or with non-neuronal cells, constituting the site of neurotransmitter release.

T

$t_{1/2}$	↪ Half-time (of association or dissociation).
Tinctorial specificity	↪ Selective ability of tissue components for colored substances called stains and used in light microscopic histology.
Total binding	↪ Total amount of measurable radioligand binding.
Transduction	↪ Transfer of information from the first to the second messenger across the cell plasma membrane.

U

Ultramicrotome	↪ Microtome for ultra-thin section making (down to 0.05 µm thickness).

V

Vibratome	↪ Vibrating microtome for making thick slices (30 to 100 µm thick) of a tissue sample that is neither frozen nor embedded in resin or paraffin.

W

Wet autoradiography	↪ Autoradiography by dipping into liquid emulsion.

Index